Environment and human well-being: a practical strategy

Lead authors
Don Melnick, Coordinator
Jeffrey McNeely, Coordinator
Yolanda Kakabadse Navarro, Coordinator
Guido Schmidt-Traub
Robin R. Sears

UN Millennium Project
Task Force on Environmental Sustainability
2005

earthscan
from Routledge

MillenniumProject

First published by Earthscan in the UK and USA in 2005

For a full list of publications please contact:

Earthscan
2 Park Square, Milton Park, Abingdon, Oxon OX14 4RN
52 Vanderbilt Avenue, New York, NY 10017

First issued in hardback 2018

Earthscan is an imprint of the Taylor & Francis Group, an informa business

A catalogue record for this book is available from the British Library

Library of Congress Cataloging-in-Publication Data

A catalog record has been requested

This publication should be cited as: UN Millennium Project. 2005. ***Environment and Human Well-being: A Practical Strategy.*** Report of the Task Force on Environmental Sustainability.

Photos: Front cover Pedro Cote/UNDP; back cover, top to bottom, Christopher Dowswell/UNDP, Pedro Cote/UNDP, Giacomo Pirozzi/Panos Pictures, Liba Taylor/Panos Pictures, Jørgen Schytte/UNDP, UN Photo Library, Giacomo Pirozzi/UNICEF, Curt Carnemark/World Bank, Pedro Cote/UNDP, Franck Charton/UNICEF, Paul Chesley/Getty Images, Ray Witlin/World Bank, Pete Turner/Getty Images.

This book was edited, designed, and produced by Communications Development Inc., Washington, D.C., and its UK design partner, Grundy & Northedge.

The UN Millennium Project was commissioned by the UN Secretary-General and sponsored by the United Nations Development Programme on behalf of the UN Development Group. The report is an independent publication that reflects the views of the members of the Task Force on Environmental Sustainability, who contributed in their personal capacities. This publication does not necessarily reflect the views of the United Nations, the United Nations Development Programme, or their Member States.

ISBN 13: 978-1-138-47182-5 (hbk)
ISBN 13: 978-1-84407-228-6 (pbk)

Foreword

The world has an unprecedented opportunity to improve the lives of billions of people by adopting practical approaches to meeting the Millennium Development Goals. At the request of the UN Secretary-General Kofi Annan, the UN Millennium Project has identified operational strategies to eradicate poverty by scaling up investments in infrastructure and human capital while promoting gender equality and environmental sustainability. These strategies are described in the UN Millennium Project's report *Investing in Development: A Practical Plan to Achieve the Millennium Development Goals* that has been coauthored by the coordinators of the UN Millennium Project task forces.

The task forces have identified the interventions and policy measures needed to achieve each of the Goals. In *Environment and Human Well-being: A Practical Strategy*, the Task Force on Environmental Sustainability highlights the critical importance of maintaining functional ecosystems, which provide food, water, fuel, climate regulation, and other essential services to the world's poorest people; curb the pollution of water and air; and mitigate climate change. The report identifies direct investments in environmental management as well as the structural policy and institutional changes required for countries to make progress towards achieving environmental sustainability.

Environment and Human Well-being proposes concrete and practical steps that governments and international agencies can undertake to operationalize the concept of environmental sustainability and to integrate the principles and practices of environmental sustainability into national development plans.

This report has been prepared by a group of leading experts who contributed in their personal capacity and volunteered their time to this important task. I am very grateful for their thorough and skilled efforts and I am sure that the practical options for action in this report will make an important contribution to achieving the Millennium Development Goals. I recommend it to anyone who is interested in how countries can achieve environmental sustainability.

Jeffrey D. Sachs
New York
January 17, 2005

Contents

Boxes

Figures

Maps

Tables

Task force members

Task force coordinators

Yolanda Kakabadse Navarro, Fundación Futuro Latinoamericano, Quito, Ecuador

Jeffrey McNeely, IUCN–The World Conservation Union, Gland, Switzerland

Don Melnick, Columbia University, New York, United States

Task force manager

Robin R. Sears, Columbia University, New York, United States

Task force members

Patricia Balvanera, Centro de Investigaciones en Ecosistemas, Universidad Nacional Autonoma de Mexico–Campus Morelia, Morelia, Mexico

David Brackett, Environment Canada, Government of Canada, Gatineau, Quebec, Canada

Damayanti Buchori, Bogor Agricultural University and PEKA Foundation, Bogor, Indonesia

Malin Falkenmark, Stockholm International Water Institute, Stockholm, Sweden

Claudia Martinez, Corporación Andina de Fomento-CAF, Caracas, Venezuela

Charles McNeill, United Nations Development Programme, New York, United States

Rodrigo Medellín, Instituto de Ecologia, Universidad Nacional Autonoma de Mexico, Distrito Federal, Mexico

Patrick Milimo, UN Millennium Project, Nairobi, Kenya

Paulo Moutinho, Instituto de Pesquisa Ambiental da Amazônia, Brasília, Brazil

Shahid Naeem, Columbia University, New York, United States

Abdoulaye Ndiaye, United Nations Development Programme, Global Environment Facility, Dakar, Senegal

Jonathan Patz, University of Wisconsin, Madison, United States

Mary Pearl, Wildlife Trust, New York, United States

Ellen Pikitch, Pew Institute for Ocean Science, University of Miami Rosenstiel School, New York, United States

Ravi Prabhu, Center for International Forestry Research, Harare, Zimbabwe

Walter V. Reid, Millennium Ecosystem Assessment, United States

Peter Johan Schei, The Fridtjof Nansen Institute, Oslo, Norway

Chikako Takase, United Nations Department of Economic and Social Affairs, New York, United States

Robert Watson, The World Bank, Washington, D.C., United States

Douglas Williamson, Food and Agriculture Organization, Rome, Italy

Kaveh Zahedi, United Nations Environment Programme—World Conservation Monitoring Center, Cambridge, United Kingdom

UN Millennium Project secretariat

Guido Schmidt-Traub, Policy Advisor, New York, United States

Preface

Humanity at the dawn of the twenty-first century faces daunting environmental challenges and choices inextricably linked to the global effort to eradicate human misery. In September 2000 world leaders at the UN Millennium Summit confirmed that pursuit of environmental sustainability is essential to poverty reduction. To move the world toward greater equity and sustainability, these leaders endorsed eight Millennium Development Goals, which focus the global community's efforts on achieving concrete and measurable objectives by the year 2015 (see Goals on pages xvi–xvii). This global framework articulates a vision of shared responsibility between rich and poor countries alike. To ensure environmental sustainability (Goal 7), the framework calls for integrating the principles of sustainable development into country policies and programs (target 9).

The Johannesburg Declaration on Sustainable Development and Plan of Implementation, announced by world leaders at the 2002 World Summit on Sustainable Development, assumes "a collective responsibility to advance and strengthen the interdependent and mutually reinforcing pillars of sustainable development—economic development, social development, and environmental protection—at the local, national, and global levels."[1] Leaders formally recognized three overarching objectives for sustainable development: eradicating poverty, changing consumption and production patterns, and protecting and managing the natural resource base for economic and social development. The task force emphasizes that meeting the last two objectives is fundamental to achieving the first: eradicating poverty over the long term.

This report translates target 9 into a call for action. It recommends needed changes in government policies, management strategies, and market dynamics at all levels to meet the Goals. General in nature, these recommendations must necessarily be tailored to the unique social, economic, and political conditions of each country. In developing its recommendations, the task force drew extensively on the Millennium Ecosystem Assessment, Intergovernmental Panel on Climate Change, and the knowledge and experience of experts around the world.

The report has two parts: Part 1, The Problem, presents the context for understanding environmental sustainability. Chapter 1 links the concept of environmental sustainability to the other Millennium Development Goals. Chapter 2 discusses the drivers of environmental degradation and diagnoses ecosystem health in six key elements of the global environment. Chapter 3 asks why the world has failed to make progress toward environmental sustainability, despite growing scientific understanding and awareness of its urgency and importance. In response, Part 2, The Solution, presents the measures required to ensure environmental sustainability. Chapter 4 outlines specific investments needed to improve environmental management. Chapter 5 describes systemic changes required to achieve environmental sustainability. Finally, Chapter 6 considers how the recommendations can be implemented at national, regional, and global levels and what roles the various actors can play.

Acknowledgments

The lead authors gratefully acknowledge the work of the other task force members, whose combined experience in the sciences, environmental management, and development provided the expertise and impetus behind much of the information and recommendations in this report. We would also like to especially thank Mary Pearl, a member of the task force, for her invaluable assistance in drafting the executive summary. At the UN Millennium Project Secretariat, Albert Cho provided important drafting support on sections of the final report, along with Alice Wiemers. We wish to thank John McArthur, UN Millennium Project manager, and Jeffrey Sachs, UN Millennium Project director, for their leadership.

The task force commissioned seven issues papers and acknowledges the authors, many of them members of the task force, for the detailed information they provided: David Brackett, Carolina Caceres, Susan Mainka, and Rodrigo Medellín on biodiversity; Patricia Balvanera and Ravi Prabhu on ecosystem services; Robert Repetto on economics; Dami Buchori, Suzana Padua, and Mary Pearl on capacity building; Jeffrey McNeely and Patrick Milimo on policy responses; Robert Watson on energy; and John Buccini and Cristina Cortinas de Nava on chemicals and pollution.

The task force as a whole also wishes to acknowledge, with deep gratitude, the numerous contributions of people and organizations, through their comments and revisions, on both the interim report and this final report. Several people at the offices of the United Nations Department of Economic and Social Affairs commented on multiple drafts; technical comments came from staff members at the World Bank and the U.K. Department for Environment, Food, and Rural Affairs and Department for International Development. Helpful comments were also made by Adnan Amin, Hussein Abaza, Charlie Arden-Clark, Thierry de Oliveira, Mark Radka, and David Smith of the

United Nations Environment Programme; Philip Dobie, Linda Ghanime, Joakim Harlin, Peter Hazlewood, Michael Hooper, Laura Lee, Maryam Niamir-Fuller, Kamal Rijal, Susanne Schmidt, Nadine Smith, Minoru Takada, Su Chin Teoh, and, especially, Gelila Terrefe of the United Nations Development Programme; and David Coates, David Cooper, and Marjo Vierros of the Secretariat for the Convention on Biological Diversity.

Specifically, on fisheries Christine Santora (Pew Institute for Ocean Science), working with task force member Ellen Pikitch, provided text and revisions. On forests, David Henderson-Howat, Pekka Patosaari, and Mia Söderlund of the Secretariat of the United Nations Forum on Forests provided input, along with task force member Ravi Prabhu's colleagues at the Center for International Forestry Research. On pollution, Cristina Cortinas, coauthor of one of the task force–commissioned background papers, provided guidance, text, and revisions.

We acknowledge the insightful comments of panelists at two public events that we held at international meetings. At the annual meeting of the Society for Conservation Biology in New York City (August 2004), René Castro (former Minister of Environment and Energy, Government of Costa Rica), Avecita Chicchón (Wildlife Conservation Society), and Larry Linden (Goldman Sachs) provided in-depth comments on the interim report. At the IUCN World Conservation Forum in Bangkok (November 2004), we received detailed comments from David Jhirad (World Resources Institute), Ian Johnson (World Bank), and Valli Moosa (former Minster of Environment, Government of South Africa) on a draft of the final report. We appreciate additional comments from Stan Bernstein and Emily White of the UN Millennium Project Secretariat, Roberto Lenton and Kristen Lewis of the Task Force on Water and Sanitation, and Calestous Juma, coordinator of the Task Force on Science, Technology, and Innovation.

The final report was edited and produced by Norma Adams, Meta de Coquereaumont, Bruce Ross-Larson, Christopher Trott, and Elaine Wilson of Communications Development Incorporated.

The institutions that generously hosted or co-organized task force meetings and provided administrative support include the Center for Environmental Research and Conservation, the Fundación Futuro Latinoamericano, the PEKA Foundation, IUCN–The World Conservation Union, and the UN Millennium Project.

In all, nearly 100 experts had direct input into the deliberations of the Task Force on Environmental Sustainability, though the members of the task force assume sole responsibility for the contents of the report.

Abbreviations

CBD	Convention on Biological Diversity
CDM	Clean Development Mechanism
CSO	civil society organization
DFID	U.K. Department for International Development
EBFM	ecosystem-based fishery management
EDF	European Development Fund
FAO	Food and Agriculture Organization (United Nations)
GEF	Global Environment Facility
GEO	Global Environment Outlook
IDA	International Development Association
IGBP	International Geosphere Biosphere Programme
IHDP	International Human Dimensions Programme
IPCC	Intergovernmental Panel on Climate Change
IUCN	The World Conservation Union
IWRM	integrated water resource management
MDG	Millennium Development Goal
NGO	nongovernmental organization
ODA	official development assistance
PRS	poverty reduction strategy
PRSP	Poverty Reduction Strategy Paper
REACH	European Commission–proposed Registration, Evaluation, and Authorization of Chemicals
SARS	Severe Acute Respiratory Syndrome
SEIA	Strategic Environmental Impact Assessments
SFM	sustainable forest management
UN	United Nations
UNCCD	United Nations Convention to Combat Desertification

UNCED	United Nations Conference on Environment and Development
UNCLOS	United Nations Convention on the Law of the Sea
UNICPOLOS	UN Informal Consultative Process on Oceans and the Law of the Sea
UNDP	United Nations Development Programme
UNEP	United Nations Environment Programme
UNESCO	United Nations Educational, Scientific and Cultural Organization
UNFCCC	United Nations Framework Convention on Climate Change
WCRP	World Climate Research Program
WHO	World Health Organization

goals

Millennium Development Goals

Goal 1

Eradicate extreme poverty and hunger

Target 1.
Halve, between 1990 and 2015, the proportion of people whose income is less than $1 a day

Target 2.
Halve, between 1990 and 2015, the proportion of people who suffer from hunger

Goal 2

Achieve universal primary education

Target 3.
Ensure that, by 2015, children everywhere, boys and girls alike, will be able to complete a full course of primary schooling

Goal 3

Promote gender equality and empower women

Target 4.
Eliminate gender disparity in primary and secondary education, preferably by 2005, and in all levels of education no later than 2015

Goal 4

Reduce child mortality

Target 5.
Reduce by two-thirds, between 1990 and 2015, the under-five mortality rate

Goal 5

Improve maternal health

Target 6.
Reduce by three-quarters, between 1990 and 2015, the maternal mortality ratio

Goal 6

Combat HIV/AIDS, malaria, and other diseases

Target 7.
Have halted by 2015 and begun to reverse the spread of HIV/AIDS

Target 8.
Have halted by 2015 and begun to reverse the incidence of malaria and other major diseases

Goal 7

Ensure environmental sustainability

Target 9.
Integrate the principles of sustainable development into country policies and programs and reverse the loss of environmental resources

Target 10.
Halve, by 2015, the proportion of people without sustainable access to safe drinking water and basic sanitation

Target 11.
Have achieved by 2020 a significant improvement in the lives of at least 100 million slum dwellers

Goal 8

Develop a global partnership for development

Target 12.
Develop further an open, rule-based, predictable, nondiscriminatory trading and financial system (includes a commitment to good governance, development, and poverty reduction—both nationally and internationally)

Target 13.
Address the special needs of the Least Developed Countries (includes tariff- and quota-free access for Least Developed Countries' exports, enhanced program of debt relief for heavily indebted poor countries [HIPCs] and cancellation of official bilateral debt, and more generous official development assistance for countries committed to poverty reduction)

Target 14.
Address the special needs of landlocked developing countries and small island developing states (through the Program of Action for the Sustainable Development of Small Island Developing States and 22nd General Assembly provisions)

Target 15.
Deal comprehensively with the debt problems of developing countries through national and international measures in order to make debt sustainable in the long term

Target 16.
In cooperation with developing countries, develop and implement strategies for decent and productive work for youth

Target 17.
In cooperation with pharmaceutical companies, provide access to affordable essential drugs in developing countries

Target 18.
In cooperation with the private sector, make available the benefits of new technologies, especially information and communications technologies

Ten key recommendations

Recommendation 1

Improve small-scale agricultural production systems.

- Increase the use of sustainable agriculture techniques to preserve natural assets by mobilizing local knowledge and experience in protecting and improving soils using agroforestry and leguminous cover crop, maintaining crop genetic diversity, and improving crop management and storage.
- Restore and manage desertified lands by developing prevention strategies to protect arid ecosystems.
- Protect surrounding natural habitat through rational land-use planning and the securing of resource and land ownership and management rights.

Recommendation 2

Promote forest management for protection and sustainable production.

- Increase real income in informal forest sector activities by at least 200 percent by harnessing the entrepreneurial spirit of informal harvesters and users of forest products, through outreach from government agencies, civil society organizations, and certification organizations; rationalizing institutional and regulatory frameworks; and creating incentives for conservation and sustainable management.
- Integrate ecosystem management of 90 percent of river basin systems—including those that span national, state, or provincial boundaries—by increasing regional coordination, providing technical assistance to land owners, and implementing best practices for natural resource use.
- Protect and restore ecologically viable areas of all major forest, shrubland, and pasture vegetation types and their biodiversity by coordinating conservation strategies, implementing a mosaic of interconnected protected areas large enough to address the threats, increasing the use of independently certified sustainable forest management practices, addressing the concerns of vulnerable populations, and compensating affected stakeholders.

Recommendation 3

Combat threats to freshwater resources and ecosystems causing water scarcity in dry areas and flooding in wet ones, and pollution and salinization.

- Reduce demand for freshwater, especially in cropping systems, by increasing water use efficiency; identify new water sources, such as rainwater and recycled wastewater; and manage demand through a supportive regulatory environment and incentives.
- Maintain pollution levels in surface water and groundwater sources below maximum allowable levels for any given pollutant by establishing and enforcing pollution targets
- Maintain aquatic biodiversity by ensuring minimum environmental flow and protect aquatic environments by rationalizing resource distribution based on the amount of water that must remain in a river system to maintain ecosystem function and by controlling alien invasive species.

Recommendation 4

Address the threats to fisheries and marine ecosystems brought on by the increasing demand for marine products and services and by the degradation of inland habitat.

- Manage fisheries sustainably by implementing ecosystem-based fishery management, applying sound research.
- Restore depleted fish populations to at least minimum target levels of biomass by eliminating unsustainable fishing practices, aligning land and water conservation policies, controlling overfishing, and establishing and achieving biomass targets.
- Establish a network of representative, fully protected marine reserves by increasing coordination and coverage of protected areas.

Recommendation 5

Address the drivers of air and water pollution.

- Reduce exposure to toxic chemicals by adopting integrated pest management strategies to reduce pesticide pollution, improving frameworks for chemicals management, and implementing standards for environmental management to produce cleaner production technologies.
- Reduce mortality and morbidity in children under five caused by pneumonia and acute respiratory infection by investing in cleaner technologies and raising public health awareness, particularly among women.
- Reduce mortality and morbidity in children under five caused by waterborne diseases by protecting water sources from untreated runoff from housing and livestock and developing new water-collection techniques.
- Reduce the atmospheric levels of the six key pollutants (carbon monoxide, lead, nitrogen dioxide, particulates, sulfur dioxide, and ozone) and methane by investing in cleaner energy technologies and improving management of organic wastes to reduce methane generation.

Recommendation 6

Mitigate global climate change by stabilizing greenhouse gas concentrations at 450–550 part per million carbon dioxide equivalent and support countries in adapting to its effects.

- Invest in the development of cost-effective no- or low-carbon energy technologies, including research and development, demonstration, and market scale-up; eliminate market failures and distortions; and internalize environmental externalities into energy prices so that environmentally friendly technologies can compete in the market.
- Promote markets for carbon by supporting multilateral instruments, such as national and international tradable emissions systems, and rationalize sustainable production and consumption patterns to mitigate carbon emissions by reducing deforestation and forest fires and increasing afforestation and reforestation.
- Operationalize responses to climate change and variability by integrating them into national economic and sector planning and investing in adaptation strategies, such as helping farmers adopt alternative cropping and water management plans in response to changing temperatures and precipitation patterns.

Recommendation 7

Strengthen institutions and governance to improve the design and implementation of strategies for achieving environmental sustainability.

- Strengthen training, recruitment, and retention of environment experts by increasing funding and salaries and improving working conditions.
- Secure sufficient funding for environmental institutions through dedicated national revenue sources and, as necessary, additional funding from donor countries.

- Reform government institutions and improve interagency coordination by streamlinin environmental and natural resource management structures and strengthening coordinating mechanisms across agencies.
- Promote good governance and improve gender equality by ensuring minimum standards of governance and the inclusion of marginalized groups in policy formatioı and decisionmaking.

Recommendation 8

Develop policy instruments to correct market failures and distortions and to align puk and private incentives with the health and well-being of the poor.

- Incorporate the cost of environmental degradation into national accounts.
- Pay for ecosystem services and reform tax systems to provide additional incentives environmental sustainability.
- Phase out environmentally harmful subsidies.
- Promote trade instruments to support the legal, sustainable harvesting of natural resource products such as timber.
- Strengthen property and land tenure rights, including community management regimes, and, where possible, formally recognize customary land management practices and indigenous people's rights to land.
- Improve national and international regulatory frameworks to contain pollution and other environmentally harmful activities.

Recommendation 9

Promote science and technology for environmental sustainability, and expand the use scientific and indigenous knowledge related to environmental management by policymakers and the general public.

- Mobilize science and technology on a national scale through direct public financing and competitions in support of sustainable development.
- Establish mechanisms for disseminating science and technology advice to policymakers at all levels of government.
- Train civil servants and political decisionmakers in environmental management throu broad-based programs that include recognition of locally derived solutions to environmental problems.
- Strengthen public comprehension of key environmental issues through public awareness campaigns, financial support of civil society organizations, and integratio of environmental sustainability issues into school curricula.
- Strengthen global scientific assessments and expand programs to review regional ar subregional environmental challenges, and initiate assessment initiatives in new areas, such as agriculture and water.

Recommendation 10

Build environmental sustainability into all development strategies across sectors, and increase funding for environmental management.

- Adopt quantified and time-bound environmental objectives to guide the design of environmental policies.
- Ensure that national poverty reduction strategies include an assessment of the environmental impacts and strategies for avoiding or mitigating potentially adverse environmental effects.
- Increase funding for national environmental programs that are integrated in poverty reduction strategies and that satisfy strict technical and operational standards, by reallocating domestic resources and raising development assistance, as needed.
- Significantly increase funding to countries to support implementation of existing multilateral environmental agreements and reporting on progress.

Executive summary

Our lives on this planet depend on nature's provision of stability and resources. Current rates of human-engendered environmental destruction threaten those resources and leave death and misery in their wake. But we can avoid this. To do so, we must act in concert and with a sense of urgency to make the structural and policy changes needed to maintain ecosystems and their services, control water and air pollution, and reverse the trends leading to global warming. This must be done if we are to achieve the level of environmental sustainability necessary to meet the UN Millennium Development Goals addressing poverty, illiteracy, hunger, discrimination against women, unsafe drinking water, and environmental degradation.

> Environmental sustainability is essential to achieving all other Millennium Development Goals..

By *environmental sustainability* we mean meeting current human needs without undermining the capacity of the environment to provide for those needs over the long term. Achieving environmental sustainability requires carefully balancing human development activities while maintaining a stable environment that predictably and regularly provides resources such as freshwater, food, clean air, wood, fisheries, and productive soils and that protects people from floods, droughts, pest infestations, and disease. Therefore, environmental sustainability is necessarily a fundamental objective in the pursuit of the seven other Millennium Development Goals. As stated in the UN Millennium Declaration, we must "spare no effort to free all of humanity, and above all our children and grandchildren, from the threat of living on a planet irredeemably spoilt by human activities, and whose resources would no longer be sufficient for their needs."

Achieving a healthy, sustainable environment first requires understanding the drivers of environmental change

Throughout this report, the Task Force on Environmental Sustainability proposes concrete measures to build development strategies on the foundation of environmental sustainability.

To understand these measures, it is first necessary to carefully and objectively analyze the state of the environment. Part 1 of this report describes the unprecedented environmental change our world has experienced over the past decade, diagnoses the root causes of environmental degradation and environmental threats to human well-being, and identifies the obstacles to achieving a healthy, sustainable environment.

With a clear understanding of the state of the environment and its relationship to human well-being, the urgent need to better manage the environment becomes evident. Actions that integrate principles of environmental sustainability into development strategies are presented in Part 2 of this report. Achieving environmental sustainability requires dramatic changes in the ways societies and citizens manage biodiversity and the wastes and byproducts of production and consumption. It also requires changes in the consumption patterns themselves. Direct investments and structural changes are required at local, national, regional, and global levels. We must also address the underlying causes of environmental problems.

The Problems

Achieving a healthy, sustainable environment first requires understanding the drivers of environmental change, assessing the state of the environment and people's dependence on it, and identifying the obstacles to ameliorating environmental degradation.

Operationalizing environmental sustainability

The pursuit of environmental sustainability is an essential part of the global effort to reduce poverty, because environmental degradation is inextricably and causally linked to problems of poverty, hunger, gender inequality, and health. Protecting and managing the natural resource base for economic and social development and changing consumption and production patterns are fundamental requirements for poverty eradication. Integrating the principles and practices of environmental sustainability into country policies and planning programs is therefore key to successful poverty reduction strategies.

Environmental sustainability issues arise at all levels, from local land use practices to global consumption and production patterns. Thus responsibility for natural resource management, waste management, and biodiversity protection must be shared among all nations. Every country must work toward environmental sustainability by defining concrete and quantifiable objectives and implementing the process targets of multilateral and regional environmental agreements.

The pursuit of environmental sustainability is an essential part of the global effort to reduce poverty

Environment and human well-being. The environment comprises a diversity of ecosystems, from forests, grasslands, and agroecosystems to freshwater systems and coral reefs. Each provides a set of benefits that contribute to human health, well-being, and livelihood, from the direct provisioning of goods to more indirect benefits, such as through regulating or supporting ecosystem services. The world's poor depend disproportionately on ecosystem services to provide for their systems of small-scale agriculture, grazing, harvesting, hunting, and fishing. Without access to infrastructure providing safe drinking water, electricity, fuel, and transportation, poor people rely on natural sources of clean air and water, fertile soil, renewable energy, and biodiversity to meet their needs. Although 1.3 billion people live on marginal lands and one-fifth of all people lack access to safe water, environmental sustainability is not adequately addressed in most countries' poverty reduction strategies.

Environment and poverty. Environmental degradation is a product of the activities of both rich and poor. Deforestation, for example, is partly caused by local demand for agricultural land or construction materials, but is even more fundamentally driven by the industrialized world's demand for timber and the growing international trade in forest products. Greenhouse gas emissions in the world's developed countries have largely driven global climate change, which threatens human well-being, ecosystem functioning, and biodiversity. If developed countries do not reduce emissions, and economic growth in newly emerging economies contributes equivalent per capita quantities of greenhouse gases, climate change will accelerate sharply.

Environment and food security. All food ultimately derives from ecosystem services. In Africa, bush meat is the main source of animal protein, while in Southeast Asia fisheries provide the main source of protein. Sustainable management of terrestrial and marine ecosystems is thus a prerequisite to global food security. Inappropriate intensive and extensive agricultural techniques cause loss of biodiversity, soil erosion, salinization of irrigated areas, agrochemical leaching, increased withdrawals of groundwater and surface water, and pesticide resistance. These forms of environmental degradation can cause irreversible losses in food availability on land and in rivers, lakes, and the sea.

Environment and health. Environmental degradation adversely affects human health through exposure to bacteria, parasites, and disease vectors (mosquitoes and snails, for example); chemical agents (such as pesticides and heavy metals); and physical and safety hazards. Gastrointestinal diseases, strongly linked to unsafe water and environmental degradation, are the leading killer of children under five. Polluted air, indoors and out, kills more than 2 million people a year. Many of today's emergent or resurgent diseases, such as encephalitis, dengue fever, and malaria, are on the rise because of human disruption of ecosystems.

If left unmanaged, the environment will continue to deteriorate

What is driving environmental change?

We are living in an era of unprecedented environmental change. In every region of the world, human actions have affected the natural environment, resulting in rapidly diminishing forests and coral reefs, increased consumption of scarce water and energy resources, desertification, spread of invasive alien species, loss of biodiversity, and increasing effects of global climate change. If left unmanaged, the environment will continue to deteriorate, impeding efforts to achieve the Millennium Development Goals.

Direct drivers. The five most significant direct drivers of environmental deterioration are:

1. *Land cover change,* resulting from logging, urbanization, conversion to agriculture, road construction, and human habitation, among other factors, can impair the delivery of vital ecosystem services, such as water retention and flood attenuation.
2. *Overappropriation or inappropriate exploitation of natural resources* can reduce even the stock of renewable resources below sustainable levels. Overfishing is an example.
3. *Invasive alien species* are non-native organisms that become established and spread in new environments. They can choke out native species, clog waterways, and threaten human health.
4. *Pollution* of air, soil, and water by chemical and organic wastes affects human health, reduces agricultural production, and damages ecosystems.
5. *Climate change* may be the single greatest driver of environmental change on a broad scale. The warming trend of the past 30–50 years has had such diverse effects as altered precipitation patterns, greater frequency of extreme weather events, rising sea levels, increased ranges for some disease vectors, and changes in ecological systems, including migration and reproduction patterns.

Indirect drivers. The causes of environmental change are complex and synergistic and include indirect as well as direct factors that lead to deterioration of ecosystems and the pollution of our air, water, and land. The task force considers the most powerful indirect drivers of environmental deterioration to include:

1. *Demographic change.* Population growth, rural-to-urban migration, and shifts in household economic status have important implications for the environment, because they tend to increase pressure on the environment.
2. *Economic factors.* Economic growth intensifies resource consumption, drives land cover change, and generates waste. But rising incomes can also bring investments in environmental improvement and cleaner technologies. Extreme poverty can drive environmental degradation, in turn reinforcing poverty.

The major indirect driver of land and soil degradation is demographic change

3. *Market failures and distortions.* Environmentally damaging subsidies can encourage overproduction or overexploitation of resources such as fisheries and forests. Failure to account for resource depletion may result in a misleading picture of economic conditions. Increased trade flows may facilitate the movement of alien species and pathogens, causing damages that further strain national accounts.
4. *Scientific and technological change.* Some new technologies can, for example, enable more effective pollution abatement, whereas other technologies might drive overexploitation by increasing resource extraction efficiency.
5. *Institutional gaps.* Malfunctioning or absent political and regulatory institutions allow overexploitation of resources, and weak enforcement regimes fail to deter damaging forms of extraction, such as illegal logging. Insufficient participation of key stakeholders in the planning and management of sustainable resource use reduces the effectiveness of policies and their implementation.
6. *Sociopolitical factors.* Differences in culture and social behavior yield varying consumption and production patterns, and social change can produce unpredictable shifts in resource use. Countries in conflict are unlikely to invest in environmental protection or other public goods.

Six key elements of the environment affecting health and economic well-being

Given these principal direct and indirect drivers of environmental change, the next step in designing appropriate interventions is to identify how these drivers influence key elements of the world's environment that affect human health and economic well-being. Six key elements are discussed in this report:

Agricultural production systems. Production systems (crops, grazing land, orchards, plantations, and freshwater aquaculture) cover almost a third of the Earth's land surface. They were carved out of natural ecosystems that formerly provided a wide range of services, including maintaining an extensive soil biota whose health is critical to land productivity and to water availability. Among the direct drivers of degradation of agricultural production systems are land cover changes, such as extensification that disrupts the soil's natural regulatory functions; inappropriate exploitation, such as inefficient irrigation; and climate change, causing such problems as more frequent droughts and flooding. The major indirect driver of land and soil degradation is demographic change—particularly population growth. Another is market distortions, driving farmers to extensify production to compensate for falling global prices for crops. Extreme poverty prevents people from investing in maintenance of soil fertility.

Climate change is profoundly affecting the function and dynamics of the world's forests

Forests. The Earth's 3.4 billion hectares of forest directly contribute to the livelihoods of 90 percent of the world's 1.2 billion people living in extreme poverty, by providing food, fuel, shelter, freshwater, fiber, bush meat, and genetic resources. Over the last decade the world has annually lost forests equivalent in extent to Portugal (9.4 million hectares).

More than 70 percent of forest destruction is driven directly by the expansion of agriculture, with urbanization, energy production, and mining contributing significantly as well. In many cases indiscriminate logging has catalyzed this destruction by providing the infrastructure for human expansion. Much of the destruction and degradation of forests could be avoided if access and tenure rights were fairer to local people, especially marginalized groups and women, who depend on forests for their livelihoods. Governments and markets have failed to understand and support the true value of forests to human development, thus opening opportunities for illegal forest activities and violent conflict. Regulations and law enforcement have been powerless to stop illegal logging, which accounts for at least half of timber extraction. Invasive alien species are another direct driver of forest degradation, and pollutants such as acid precipitation, tropospheric ozone, and elevated levels of carbon threaten forests by changing physiological processes and altering the behavior of plant pests. Climate change is profoundly affecting the function and dynamics of the world's forests. Indirect drivers of forest habitat deterioration include demographic changes, economic factors, and institutional gaps.

Freshwater resources and ecosystems. Freshwater resources are fundamental to human survival. They support life by enabling food and energy production and serving as a transport medium. In addition, wetlands, lakes, and rivers mitigate floods. The current per capita availability of water varies considerably globally, but overall has fallen by half during the past 40 years, and more than half the world's natural wetlands have disappeared. Irrigated agriculture accounts for 70 percent of water withdrawals worldwide, and about a third of water use depends on unsustainable withdrawals, mostly in Asia, the Middle East, and Africa.

Direct drivers of degradation include dams for energy generation, channelization and flow diversion for irrigation and flood control, wetland drainage, and groundwater withdrawal. Overappropriation of water takes the form of excessive water diversion for irrigation and urban use, which create shortages and lead to salinization. Biological pollution of water is responsible for 2.2 million deaths a year. Chemical pollution also jeopardizes human health. Climate changes in the form of increased droughts and floods affect surface water availability for human needs. Major indirect drivers of freshwater deterioration are demographic change, economic factors, and institutional gaps. Population growth and urbanization increase demand. Inappropriate water pricing policies and agricultural subsidies undervalue water and weaken incentives to manage it sustainably, and poorly integrated management across sectors causes shortages, pollution, conflict, and inefficiently allocated claims on limited resources.

Direct drivers of the deterioration of fisheries and marine ecosystems include over-appropriation and pollution

Fisheries and marine ecosystems. Oceans cover about 70 percent of the planet's surface and are by far the largest habitat for life on Earth. They supply billions of people with food and mineral resources. Fisheries supply 16.5 percent of the animal protein consumed worldwide—significant considering that undernutrition is the major cause of human mortality, accounting for 30 percent of deaths. Some 75 percent of all wild fish stocks are overexploited. Although coral reefs provide fish and seafood for a billion people in Asia alone, 80 percent of reefs are at risk from coastal development, fishing-related pressures, and climate change. Marine biodiversity provides critical services to our planet through climate control, carbon sequestration, and oxygen generation. Coastal waters support tourism and recreation as well as fisheries.

Direct drivers of the deterioration of fisheries and marine ecosystems include overappropriation—destructive and nonselective fishing practices, which damage ecosystems, massively deplete fish stocks, and pose a major threat to biodiversity. Other important direct drivers are pollution, from ocean dumping and downstream effects of land-based activities; invasive alien species introduced to estuaries and bays by the exchange of ballast water (some 3,000 species each day); and climate change, with rising sea-surface temperatures affecting the sea level, ice cover, salinity, and ocean circulation, and leading to coral reef bleaching. Indirect drivers are demographic change, economic factors, and institutional gaps, through increased demand arising from population growth, damaging subsidies that cause overexploitation of fisheries, and poor policies and enforcement coupled with insufficient attention to scientific advice. Taken together, these drivers are resulting in the global and local collapse of key marine populations.

Air and water pollution. Clean water and air are preconditions for human life and healthy ecosystems. Pollutants (such as carbon monoxide, lead, nitrogen dioxide, particulates, sulfur dioxide, and ozone) can cause brain damage, respiratory illness, cancer, endocrine disorders, and even death. In cities in developing countries, ambient levels of such pollutants often exceed World Health Organization guidelines. Indoor air pollution generated by cooking fuels causes respiratory disease, birth defects, and other illnesses among the estimated 2.4 billion people who burn biomass and coal in their homes, killing nearly 2.5 million children each year. Marine and freshwater pollution from household, industrial, and biological waste causes coral reef bleaching, eutrophication, and bioaccumulation of toxic substances in marine and freshwater animals. Over half the world's major rivers and associated lakes, wetlands, and groundwater areas are contaminated by pollutants from untreated sewage, chemical discharge, petroleum leaks and spills, mining residue, and run-off of sediment and nutrients from agricultural fields.

Attempts to improve air and water quality are frustrated by weak regulatory and enforcement regimes

Direct drivers of water pollution include land cover change, with new agricultural development resulting in the transmission of organic compounds, chemical fertilizers, and pesticides into bodies of water. Overappropriation of fossil fuels for expanding vehicle fleets significantly increases airborne pollutants, especially where leaded fuel is permitted. Indirect drivers are demographic change, economic factors, institutional gaps, and sociopolitical factors. Rural-to-urban migration has increased demand for energy and transportation; local economic realities thwart pollution prevention, mitigation, and abatement technologies; and attempts to improve air and water quality are frustrated by weak regulatory and enforcement regimes.

Global climate change. A stable climate provides the critical regulating services on which all ecosystems depend, affecting weather, human health, agricultural and marine productivity, the distribution and health of species, and energy consumption (for heating or cooling technologies). Since the advent of the industrial era (around 1750), greenhouse gases have increased substantially. Carbon dioxide, for example, has risen from 280 to 370 parts per million over that time and is projected to increase to 540–970 parts per million by 2100 because of energy and land-use practices, even without accounting for additional climate-induced releases of carbon dioxide from the biosphere. Industrial countries currently emit about 10 times more carbon per capita than do developing countries.

Human activities are causing climate change, and although industrial countries are primarily responsible, developing countries and poor people are the most vulnerable. The global mean surface temperature has increased over the past century, with the 1990s being the warmest decade on record. Temporal and spatial patterns of precipitation have changed, and the sea level has risen 10–25 centimeters. These trends are expected to continue, and the variability of climate patterns and the incidence of extreme weather events are projected to increase, resulting in the inundation of low-lying small island developing states and deltaic regions, particularly of countries in the South Pacific and the Indian Ocean. Tens of millions of people are expected to be displaced, and rates of malaria, dengue, and other diseases are expected to rise. Other anticipated effects include lower agricultural productivity and greater water scarcity and ecosystem disturbances. Unless mitigated, climate change may undermine efforts to achieve the Millennium Development Goals.

Direct drivers of climate change include pollution through fossil fuel combustion and land cover change, primarily tropical deforestation. The indirect drivers are demographic change, economic factors, market failures and distortions, scientific and technological change, institutional gaps, and sociopolitical factors. Most are related to the demand for energy and changes in land use, which affect emissions of greenhouse gases and result in turn in climate change.

Since the 1972 Stockholm Conference most environmental trends have worsened

Immediate regional concerns

While all regions face the global problems of climate change, biodiversity loss, and fisheries decline, each region faces distinct immediate concerns, and overall progress toward environmental sustainability varies considerably. In Latin America, home to half the world's species of plants and animals, the most pressing issues are deforestation, pollution, and damage to coastal and marine ecosystems. In small island developing states, including Caribbean and Pacific islands, key problems are climate change, marine ecosystem health, alien invasive species, and pollution. In Sub-Saharan Africa, the major environmental issues are soil and land degradation, depletion of forests and freshwater resources, and poor indoor air quality. The Middle East and North Africa suffer most from declining per capita water resources, loss of arable land, pollution-related health problems, and weak environmental institutions and legal frameworks. South Asia's most pressing environmental problems are freshwater scarcity and pollution, and soil and land degradation, while Central Asia's major environmental challenges involve land cover change and freshwater degradation. Finally, East and Southeast Asia suffer mostly from soil and land degradation, deforestation, and poor urban air quality.

Why environmental sustainability is elusive: eight obstacles to ameliorating environmental degradation

Since the 1972 Stockholm Conference on the Human Environment first focused international attention on environmental degradation, most environmental trends have worsened, despite three decades of political arrangements, high-level pronouncements, public exhortations, and over a dozen major multilateral environmental agreements.

There have been some improvements. The Montreal Protocol has successfully curbed emissions of ozone-depleting substances, many countries have improved air and water quality, and large shares of land ecosystems have been placed under protection. Yet most regions are not on track to halt environmental degradation, and some have even experienced dramatic declines in environmental quality, despite the existence of tools and policies to attenuate or mitigate direct and indirect drivers of environmental change.

The task force identifies and describes eight factors it considers obstacles to the enactment of policies and actions known to ameliorate environmental degradation:

- *Lack of clear operational objectives.* If environmental strategies cannot be monitored, they become difficult to implement.
- *Insufficient direct investment in environmental management.* In most low-income and some middle-income countries, domestic resources may be insufficient to cover the full range of needed investments in social services, infrastructure, and improved environmental management.

The task force recommends taking an ecosystem-based approach to environmental management

- *Poor integration into sector policies.* Too often, national environmental strategies are developed in isolation from other sectoral strategies that strongly affect environmental sustainability.
- *Inadequate institutional capacity, misalignment of goals, and poor governance.* In most countries environmental ministries lack the technical expertise and equipment needed to conduct effective work, follow strategies that are not congruent with those of other ministries, and are often impeded by poor governance, high-level corruption, and failures of enforcement.
- *Widespread market failures and distortions.* Market failures can create environmental degradation by misaligning the incentives of individuals or corporations with the interests of society at large.
- *Underinvestment in science and technology.* Science and technology are critical for achieving environmental sustainability by improving the understanding and monitoring of the state of the environment and by developing means to mitigate environmental degradation. However, most developing countries do not invest sufficiently in science and technology.
- *Difficulty of regional and international cooperation.* National environmental policies and investments have been stalled by inadequate regional and global management and cooperation.
- *Limited public awareness.* The public's awareness of the consequences of human-induced changes to the environment is insufficient to build a broad-based constituency for environmental sustainability.

Suggested solutions

Achieving environmental sustainability requires dramatic changes in the ways societies and citizens manage biodiversity and the processes of production and consumption. Direct investments in environmental management and structural changes are required at local, national, regional, and global levels to address the underlying causes of environmental problems.

Investing in environmental management: a call for action

To integrate environmental sustainability into national development strategies, the task force recommends taking an ecosystem-based approach to environmental management, integrating social, ecological and economic concerns in strategies for securing the restoration and sustainable use of fisheries, freshwater, forests, and soils, coupled with actions to reduce the production of chemical and organic wastes, and greenhouse gases.

The ecosystem approach, the framework for implementation adopted by the Convention on Biological Diversity, emphasizes the links between human activity and the sustainable functioning of ecosystems at a full range of spatial scales. Inherent in this approach are three major conditions to achieve long-term

The ecosystem approach emphasizes the links between human activity and the functioning of ecosystems

environmental sustainability: Biodiversity conservation and environmental management must not be confined to protected areas. Solutions must be integrated across sectors so that advances in one sector do not preclude advances in others. And investments must be implemented in an enabling environment of sound institutions and policies, market equity for the poor and women, and access to information and technologies for all.

Within an ecosystems-based approach, the task force recommends direct investments in integrated environmental management in six key elements of the environment (box 1).

- *Agricultural production systems.* Around the world agricultural systems are increasingly vulnerable to overuse, inappropriate practices, and altered weather patterns. The task force recommends increasing the use of sustainable agriculture techniques to preserve natural assets, restoring and managing desertified lands, and protecting surrounding natural habitat.
- *Forests.* To confront the negative drivers of land clearing for agriculture and logging, as well as pollution and global climate change, the task force recommends increasing real income in informal forest sector activities by at least 200 percent by harnessing the entrepreneurial spirit of harvesters of forest products, integrating ecosystem management of 90 percent of river basin systems, and protecting and restoring ecologically viable representative areas of all major forest, shrubland, and pasture vegetation types and their biodiversity.
- *Freshwater resources and ecosystems.* Increasing water scarcity in dry areas and flooding in wet ones, exacerbated by climate change, threaten household subsistence activities, agriculture, and aquatic ecosystems. Pollution and salinization pose risks to irrigated agriculture and human and wildlife health. To combat these drivers, the task force recommends reducing demand, especially in cropping systems; maintaining pollution levels in surface water and groundwater sources below maximum allowable levels for individual pollutants; and maintaining aquatic biodiversity by ensuring minimum environmental flow and protecting aquatic environments.
- *Fisheries and marine ecosystems.* Increasing demand for marine products and services, coupled with degradation of inland habitat, is resulting in irreversible losses in fish stocks, coral reefs, and the productivity of all aquatic ecosystems. The task force recommends managing fisheries sustainably, restoring depleted fish population levels, and establishing a network of representative, fully protected marine reserves.
- *Air and water pollution.* Current energy practices, mismanagement of toxic chemicals, and the conversion of natural habitats and related patterns of overproduction, overconsumption, and mismanagement of ecosystems have resulted in unsustainable levels of air and water pollutants.

Box 1

Task force recommendations for direct investments in integrated environmental management in six key elements of the environment

Agricultural production systems

Increase the use of sustainable agriculture techniques to preserve natural assets:

- Protect and improve soils, including enhanced carbon sequestration.
- Use water sustainably.
- Maintain crop genetic diversity.
- Mobilize local knowledge and experience.
- Improve crop research, management storage, and use.

Restore and manage desertified lands:

- Adopt prevention strategies to protect arid ecosystems.
- Mobilize information and technology.

Protect surrounding natural habitat:

- Rationalize land-use planning.
- Set up systems of communal ownership and management rights.

Forests

Increase real income in informal forest sector activities by at least 200 percent, by harnessing and channeling the entrepreneurial spirit of harvesters of forest products, illegal loggers, pit sawyers, wood carvers, bush meat hunters, and traders:

- Provide outreach to informal users from government agencies, civil society organizations, and certification organizations.
- Rationalize institutional and regulatory frameworks.
- Create incentives for conservation and sustainable management.

Integrate ecosystem management of 90 percent of river basin systems—including those that span national, state, or provincial boundaries:

- Increase regional coordination.
- Provide technical assistance.
- Implement best practices for natural resource use.

Protect and restore ecologically viable representative areas of all major forest, shrubland, and pasture vegetation types and their biodiversity:

- Coordinate conservation strategies.
- Implement a mosaic of interconnected protected areas of a size commensurate with addressing the threats, such as climate change.
- Increase use of independently certified sustainable forest management practices.
- Address concerns of vulnerable populations.
- Compensate affected stakeholders.

Freshwater resources and ecosystems

Reduce demand, especially in cropping systems:

- Increase water-use efficiency.
- Identify new water sources, such as rainwater and recycled wastewater.
- Manage demand through a supportive regulatory environment and incentives.

Maintain pollution levels in surface water and groundwater sources below maximum allowable levels for any given pollutant:

- Establish and enforce pollution targets.

Maintain aquatic biodiversity by ensuring minimum environmental flow and protecting aquatic environments:

- Rationalize resource distribution by determining the amount of water that must remain in a river system to maintain ecosystem function.
- Control alien invasive species.

Fisheries and marine ecosystems

Manage fisheries sustainably:

- Implement ecosystem-based fishery management based on sound research.

Box 1

Task force recommendations for direct investments in integrated environmental management in six key elements of the environment

(continued)

Restore depleted fish population levels:

- Eliminate unsustainable fishing practices.
- Align land and water conservation policies.
- Control overfishing.
- Establish and achieve biomass targets in order to restore depleted fish populations to at least minimum target levels of biomass.

Establish networks of representative, fully protected marine reserves:

- Increase coordination and coverage of protected areas.

Air and water pollution

Reduce exposure to toxic chemicals by vulnerable groups:

- Adopt integrated pest management strategies to reduce pesticide pollution.
- Improve frameworks for chemicals management.
- Implement standards for environmental management to produce cleaner production technologies.

Reduce substantially the under-five mortality and morbidity rates caused by pneumonia and acute respiratory infection:

- Invest in cleaner technologies.
- Raise public health awareness, particularly among women.

Reduce substantially the under-five mortality and morbidity rate caused by waterborne diseases:

- Protect water sources from untreated runoff from housing and livestock.
- Develop new water-collection techniques.

Reduce the atmospheric levels of the six key pollutants (carbon monoxide, lead, nitrogen dioxide, particulates, sulfur dioxide, and ozone) and methane:

- Invest in cleaner energy technologies.
- Improve management of organic wastes to reduce methane generation.

Global climate change

Invest in cost-effective and environmentally sustainable energy:

- Develop no- or low-carbon energy technologies, including research and development, demonstration, and market scale-up.
- Eliminate market failures and distortions, such as fossil fuel subsidies, and internalize environmental externalities into energy prices so that environmentally friendly technologies can compete in the market.

Promote and engage climate-friendly carbon and technology markets:

- Support multilateral instruments, such as national and international tradable emissions systems.
- Establish and commit to long-term stabilization targets for atmospheric concentrations of greenhouse gases (450–550 parts per million of carbon dioxide equivalent, for example).
- Rationalize sustainable production and consumption patterns to mitigate carbon emissions, by reducing deforestation and burning and increasing afforestation and reforestation.

Take adaptation measures:

- Mainstream responses, integrating issues of climate change and variability into national economic and sector planning.
- Invest in adaptation strategies, such as helping farmers adopt alternative cropping and water management strategies in response to changing temperatures and precipitation patterns.

Any inability to make structural changes will impede progress on achieving environmental sustainability

The task force recommends reducing vulnerable groups' exposure to toxic chemicals; reducing by 20 percent the under-five mortality and morbidity rates caused by pneumonia, acute respiratory infection, and waterborne diseases; and reducing the atmospheric levels of the six key pollutants (carbon monoxide, lead, nitrogen dioxide, particulates, sulfur dioxide, and ozone) and methane.

- *Global climate change.* Human activities such as fossil fuel combustion and deforestation, by releasing greenhouse gases, are changing the Earth's climate. To mitigate the anticipated increase in extreme weather events and the rise in sea level, the task force recommends investing in cost-effective and environmentally sustainable energy, promoting and engaging climate-friendly carbon and technology markets, and taking adaptation measures.

Four structural changes for environmental sustainability

While absolutely necessary, direct investments in environmental management will not succeed unless major structural changes are made to policies at the national, regional, and global levels. These changes are highly charged politically, but any inability to make them will stand as a major impediment to achieving environmental sustainability.

The task force recommends four fundamental changes in the way institutions and economies operate at national, regional, and global scales, so that countries can effectively integrate environmental concerns into all development and sector policies (box 2).

To take action toward achieving environmental sustainability, countries must first reform and strengthen environmental institutions and governance. This will require building up the environmental expertise available within agencies and ministries for agriculture, energy, transport, water supply, and others. Such institutional strengthening requires increased investments, particularly in human resources, and reform of institutional arrangements, including management systems.

The second fundamental structural change is the development of policy instruments at all levels to correct market failures and distortions in ways that realign public and private sector incentives with the health and well-being of ordinary citizens. The most important policy instruments to address market failures are those that incorporate the cost of environmental degradation in national accounts, introduce payments for ecosystem services and tax reform, phase out environmentally harmful subsidies, develop trade regulations, strengthen property and land-tenure rights, and improve the regulatory framework.

The third necessary structural change is to mobilize science and technology for sustainable development and to improve access to scientific and technical knowledge. Research institutions can provide technical solutions for some

Box 2
Task force recommendations for structural changes

Strengthen institutions and governance.

- Train, recruit, and retain environment experts.
- Secure sufficient funding for environmental institutions.
- Reform government institutions and improve interagency coordination.
- Improve governance and gender equality.

Correct market failures and distortions.

- Account for the cost of environmental degradation in national accounts.
- Introduce payments for ecosystem services.
- Reform tax structures to promote environmentally beneficial actions.
- Phase out environmentally harmful subsidies.
- Develop trade regulations to promote legal, sustainable harvesting of natural resource products such as timber.
- Strengthen property and land tenure rights, including community management regimes.
- Improve national and international regulatory frameworks.

Promote science and technology for environmental sustainability.

- Mobilize science and technology for sustainable development.
- Establish mechanisms for science and technology advice.
- Train decisionmakers for environmental management.
- Improve extension training and services.
- Expand education in science, mathematics, and environmental studies.
- Provide public access to environmental information.
- Strengthen global scientific assessments.

Build environmental sustainability into all development strategies across sectors.

- Adopt quantified and time-bound environmental objectives.
- Incorporate environmental sustainability into poverty reduction strategies.
- Increase funding for national environmental programs.
- Increase funding to countries to support implementation of existing multilateral environmental agreements.

of the problems of environmental degradation. Countries must gather data and develop indicators to understand current conditions and must be able to translate data into actionable information to guide environmental decisionmaking by policymakers and the general public. While the needed skills, capabilities, and information products differ in each case, the overarching principle is constant: better information and greater knowledge capacity can significantly improve the quality of decisions and their environmental outcomes.

Finally, countries must adopt quantified and time-bound environmental objectives to guide the design of environmental policies. National poverty reduction strategies should reflect an assessment of potential environmental impacts and development of strategies for avoiding or mitigating them. Countries will require increased funding for national environmental programs that are integrated in poverty reduction strategies.

It is in the implementation of recommendations and actions that there is the greatest need for innovation

Mainstreaming environmental sustainability: innovations in implementation

Many of the management actions and structural changes recommended by the task force are not new: they can be found in the major multilateral environmental agreements of the past three decades. But despite this long history of calls for action, most recommendations have not been systematically implemented. The task force, therefore, believes that it is in the implementation of recommendations and agreed-upon actions that there is the greatest need for innovation.

While environmental challenges act at local, national, regional, and global levels, this report emphasizes the need for national-level implementation to lay the foundation for successful regional and global implementation and to make the rapid progress needed to achieve the Millennium Development Goals. But because of the global context in which all countries have to work today, it is also necessary to establish a global framework of rules, regulations, and incentives.

Implementation must also bring together expertise from a broad range of fields, including representatives from most line ministries, civil society organizations, local communities, and natural and social scientists. A major implementation challenge is to identify, evaluate, and address tradeoffs and synergies between sectoral strategies and environmental objectives. Finally, implementation mechanisms must address long-term environmental change, as well as short- or medium-term social and economic imperatives. For example, a challenge such as increasing the water efficiency of agriculture requires strategies and investments that must be pursued over many years if they are to show results. An investment to reduce greenhouse gas emissions imposes a cost on today's generation, with most of the benefits of a more stable climate accruing at some future time. No tested institutional models yet exist to deal effectively with intergenerational tradeoffs, even though they lie at the heart of the implementation challenge.

National implementation mechanisms

Poverty reduction strategies, common in many low-income countries, tend not to address environmental sustainability in a systemic way. Those that do, focus only on providing access to water supply and sanitation, ignoring forest management, prevention of land degradation and desertification, pollution reduction, or other key elements outlined in this report. Poverty reduction strategies could be aligned with the Millennium Development Goals (MDGs); by covering the long term, rather than a short-term, three-year horizon; by integrating sectoral approaches to issues such as environment, gender equality, and urbanization; and by being adequately funded.

Integrated development framework. The task force endorses the UN Millennium Project recommendation that countries develop a single integrated development framework to meet all Millennium Development Goals. This MDG-based

The task force recommends that each country convene an MDG planning group

poverty reduction strategy can be developed on the basis of existing Poverty Reduction Strategy Papers (PRSPs), national strategies for sustainable development, or another framework that provides an integrated operational plan for implementing and financing strategies to achieve all the Goals and ensure appropriate monitoring of progress over a 10-year horizon.

MDG planning group. The task force recommends that each country convene an MDG planning group, chaired by the national government and including all stakeholders—donors, UN agencies, provincial and local authorities, and civil society leaders, including indigenous and women's organizations. This planning group can organize working groups, each with broad participation, for areas including health, rural development, and environmental sustainability. Each working group would include environment and gender equality experts. The environmental working group would be tasked with operationalizing Millennium Development Goals target 9 (integrate the principles of sustainable development into country policies and programs and reverse the loss of environmental resources) by establishing specific time-bound targets plus indicators for monitoring; identifying policy and investment needs to improve the management of environmental resources; and appraising the environmental effects of strategies proposed by other thematic working groups.

Operational targets. The task force further recommends that countries adopt operational targets for the environment in a process initiated by the environmental working group but inclusive of the highest level of decisionmaking, with all relevant line ministries and key stakeholders, including nongovernmental organizations, religious groups, and the commercial sector. Drawing on the best available science, the target selection team should begin with a long list of potential targets, treated as options with cost-benefit information for each. These preliminary targets should be reviewed by the MDG planning group and approved after an open discussion of tradeoffs between environment objectives and exigencies in other sectors, so that a common position on the environmental objectives is reached across all working groups. Once adopted, the environment objectives should guide the work of all thematic working groups, and each sector strategy should be carefully reviewed to assess its compatibility with the environmental objectives.

Needs assessment. In a second step, the working group on environmental sustainability would oversee the preparation of a detailed needs assessment to quantify the human and financial resources needed to meet each environmental objective agreed to by the MDG planning group. In many countries this needs-based approach will mark an important departure from current practice, which focuses primarily on the marginal expansion of services and investments, with little regard for medium- and long-term objectives. To the

A viable financing strategy is required in all countries

knowledge of the task force, no national MDG needs assessment has been conducted for the environment. Industrial countries and international organizations should offer technical assistance to developing countries that wish to carry out such assessments.

Financing strategy. Investment needs in environmental sustainability are likely to be high. A viable financing strategy for achieving environmental sustainability is required in all countries but is currently lacking in most. This report recommends making more funding available by reallocating resources to the environment, increasing domestic resource mobilization, and raising levels of development assistance and private sector investment where needed.

Monitoring. Tied to financing is another imperative: monitoring progress toward achieving environmental sustainability goals. Indicators of progress are needed at global, regional, and national levels. At the global level these indicators could be linked with the Millennium Development Goals, the 2010 biodiversity targets, and other globally accepted goals. At the national level the indicators could be linked to the poverty reduction strategies developed for national and local actions. Countries must generate relevant data and develop indicators that help them frame and implement policies and monitor policy success. Many countries need technical and financial assistance to build capacity in primary data collection, data processing and management, and development of integrated databases and information monitoring systems. Most countries do not systematically monitor key environmental parameters such as air and water quality, biodiversity, or land degradation, let alone develop indicators from these data for use by decisionmakers and the public. MDG-based poverty reduction strategies must identify the investments needed to collect data and develop and monitor indicators.

Regional and global implementation mechanisms

While the task force has focused on the country scale, regional and global scales are also important. At the regional level, mechanisms for improved environmental management are neglected and must be strengthened to support country-level MDG-based poverty reduction strategies. Some challenges, such as climate change, fisheries decline, illegal trade in forest and wildlife products, and ozone depletion, can be managed only through global implementation mechanisms.

To address such issues, the task force recommends setting up a coordinating mechanism for conventions and agreements related to environmental sustainability—such as the Convention on Biological Diversity, the Convention to Combat Desertification, the United Nations Framework Convention on Climate Change, the Ramsar Convention on Wetlands, the United Nations Convention on the Law of the Sea, and those on chemical issues—to develop

> **The long-term success in meeting all of the Millennium Development Goals depends on environmental sustainability**

joint programs to find synergies and identify tradeoffs among actions taken under these conventions. A periodic synthesis of the findings of international assessments, including the Millennium Ecosystem Assessment, Intergovernmental Panel on Climate Change, and International Assessment of Agricultural Science and Technology for Development, would likely reveal such synergies and tradeoffs.

Many international agreements have excellent work plans and long lists of priority actions. Yet few have been implemented. The task force believes that here again innovations in implementation may be useful and recommends that conventions focus on effectively supporting national interventions to stem environmental degradation, mostly by making scientific knowledge and operational best practices available to countries. The task force recommends that conventions strengthen their operational expertise and scientific capacity and focus on the enforcement, implementation, and design of national programs.

Conclusion

Environmental challenges are both complex and unique. Many institutions must act in concert to respond to them, and proposed solutions must be adapted to regional and local conditions. For this reason, this report provides neither a blueprint for achieving Goal 7 nor quantitative targets for every problem. Rather, the task force offers recommendations for how to organize the process of integrating the principles of environmental sustainability into all policies and management strategies. It draws as much attention to implementation challenges as to needed management actions and structural changes. Neither structural changes nor technical interventions will succeed unless strong support for these changes comes from national governments, nongovernmental organizations, an informed citizenry, and the larger, multilateral community.

The long-term success in meeting all of the Millennium Development Goals depends on environmental sustainability. Without it, gains will be transitory and inequitable. The paramount importance and clear urgency of environmental sustainability dictates immediate actions at all scales—and the political, social, and financial will necessary to sustain those actions.

1 The problem

The problem

The need to protect and maintain the natural environment is frequently pitted against development objectives as if the two were fundamentally incompatible. Nothing could be farther from the truth. Environmental sustainability, including the conservation of biodiversity, underpins human well-being. It is essential to the achievement of the other Millennium Development Goals and must therefore be integrated into development strategies across all sectors. But the Task Force on Environmental Sustainability maintains equally firmly that environmental strategies must be guided by the public and private human benefits derived from a healthy and functioning environment and that this is particularly salient for people living in poverty. Conservation strategies that pay insufficient attention to the immediate needs of people are doomed to fail, as are development strategies that fail to take into account the social and human costs of environmental degradation.

Achieving environmental sustainability therefore requires carefully balancing human needs with maintaining functioning ecosystems and curbing the pollution of water and air. At times, human well-being and environmental concerns are perfectly aligned, giving rise to win-win opportunities. More often, however tradeoffs will need to be carefully managed. This report provides practical guidance on how to design strategies to maintain an appropriate balance between conservation and resource use as a means toward environmentally sustainable development.

The task force proposes concrete measures to build development strategies on the foundation of environmental sustainability. Part 1 describes the unprecedented environmental change witnessed over the past decade, diagnoses the root causes of environmental degradation, and identifies the obstacles to achieving environmental sustainability. Building on this analysis, Part 2 lays out practical approaches to strengthen the management of the environment, integrate the principles of environmental sustainability into development strategies, and address the indirect drivers of environmental change.

Chapter 1

Call for operationalizing environmental sustainability

The pursuit of environmental sustainability is an essential part of the global effort to reduce poverty. This was confirmed at the turn of the millennium in two important declarations. In September 2000, world leaders agreed to the Millennium Development Goals at the United Nations Millennium Summit. Highlighting the stark choices facing humanity at the dawn of the twenty-first century, the eight Goals and their 18 associated targets focus efforts of the world community on eradicating poverty in all its forms by achieving concrete and measurable objectives by the year 2015, thereby moving the world toward greater equity and sustainability (see the Goals on pages xvi–xvii). This global framework for action articulates a vision of shared responsibility between rich and poor countries to meet time-bound, quantified objectives. Among these objectives is to ensure environmental sustainability in development planning and implementation, formally stated in Millennium Development Goal 7.

At the 2002 World Summit on Sustainable Development, world leaders adopted the Johannesburg Declaration on Sustainable Development and Plan of Implementation. This declaration assumed "a collective responsibility to advance and strengthen the interdependent and mutually reinforcing pillars of sustainable development—economic development, social development and environmental protection—at the local, national, regional and global levels."[1] Leaders also formally recognized that poverty eradication, changing consumption and production patterns, and protecting and managing the natural resource base for economic and social development are three overarching objectives for sustainable development and essential requirements of it. This report emphasizes that meeting the last two objectives are fundamental requirements for meeting the first: eradicating poverty over the long term.

This report translates the call for environmental sustainability in Goal 7 into actionable recommendations for changes in policies, management strategies, and market dynamics that must be implemented at the local, national, regional and international levels. It focuses on priority interventions needed to meet the Goals. The task force recognizes that its recommendations, while general in nature, must necessarily be tailored by the implementing agencies to the context of each country and its unique environmental challenges. In developing these recommendations, the task force has drawn extensively on the work of the Millennium Ecosystem Assessment[2] and the Intergovernmental Panel on Climate Change, as well as the knowledge and experience of experts around the world.

Operationalizing environmental sustainability is complex

Environmental sustainability can be defined as meeting human needs without undermining the capacity of the environment to provide for those needs and support life over the long term. However, this concept is difficult to operationalize for many reasons. Primary among them is the absence of specific national and international outcome targets. The Johannesburg Declaration on Sustainable Development and Plan of Implementation stated that "to reverse the current trend in natural resource degradation as soon as possible, it is necessary to implement strategies which should include targets adopted at the national and, where appropriate, regional levels to protect ecosystems and to achieve integrated management of land, water, and living resources, while strengthening regional, national, and local capacities."

The Johannesburg Plan of Action and the most up-to-date decisions under some of the major international environmental conventions set targets such as the significant reduction of the rate of loss of biodiversity by 2010 under the Johannesburg Plan of Action, specific targets in the Plant Conservation Strategy of the Convention on Biological Diversity, and the adoption of strategies for integrated water resources management by 2005. While important, such agreements lack specific quantitative outcome targets, making it impossible to track progress and evaluate implementation success. This limitation, in turn, hinders learning, ongoing improvement of environmental strategies and plans, and justification of substantial investments in environmental management.

Millennium Development Goal 7 contrasts sharply with other Goals that set concrete benchmarks to be achieved by 2015. For example, Goal 4 has a clear target: reduce child mortality by two-thirds between 1990 and 2015. With the help of epidemiology, medical science, and decades of experience in providing specific medical interventions, each country can develop the range of necessary interventions, means for their implementation, and the interim milestones to reach the goal by 2015, barring major cataclysms, such as the earthquake and tsunamis of December 2004, which claimed the lives of so many children. Indeed, best practices are well documented and available to guide national policymakers.

Operationalizing environmental sustainability is more complex. Variability of species, ecosystems, and physical processes over time and space renders "the environment" a major management challenge. Moreover, environmental and

Environmental degradation is linked to poverty, hunger, gender inequality, and poor health

social objectives imply many tradeoffs. For example, establishing a protected area to conserve biodiversity often creates conflict with local residents, who use nearby resources to earn their livelihood. Similarly, reconciling competing demands on scarce environmental resources with the need to protect vital ecosystems is an ongoing challenge. Finally, arbitrating the interests of present and future generations—the objective of sustainability—is no easy task.

To move beyond such debates and obstacles, this task force offers the following implications of Millennium Development Goal 7 and its associated target 9: Environmental degradation is a major development issue that is inextricably and causally linked to the problems of poverty, hunger, gender inequality, and health, among others. A healthy, functional natural environment—a sustainable environment—provides many of the necessary conditions for achieving the Goals and in some cases is the foundation for reaching them. Traditionally, however, many national governments have assigned environmental concerns a low priority; the environment has often taken a backseat to seemingly more urgent problems of transportation and energy, for example.

It thus becomes fundamental to integrate the principles and practices of environmental sustainability into country policies and planning programs. Since environmental sustainability extends beyond local land-use practices to global consumption and production patterns, responsibility for working toward it must be shared. It is imperative that middle-income and wealthy nations, as well as poor ones, work together to make significant changes to national, regional, and global systems and institutions related to natural resource management, waste management, and biodiversity protection. These measures, among others, are necessary to achieve and sustain a healthy environment, thereby improving the health and well-being of all people.

Fundamental to environmental sustainability is management of the supply of those resources and waste sinks (Pearce and Barbier 2000). Thus, all countries must work toward environmental sustainability by identifying concrete and quantifiable objectives and implementing the process targets of the many existing multilateral and regional environmental agreements. This chapter makes the case for doing so.

Environment and human well-being

For too many of the world's people, environmental degradation eclipses the hopes of meeting even the most basic human needs. In developing countries, one person in five lacks access to safe water, 1.0 billion people live in drylands damaged by soil degradation, and 1.2 billion live on less than $1 a day. All face the effects of a degraded natural environment.

The environment comprises a diversity of ecosystems—from forests, grasslands, and agroecosystems to freshwater systems and coral reefs. Each provides a suite of provisioning (goods), regulating, cultural, and supporting services—all of which contribute to human health, well-being, and livelihood

The value of ecosystem services is not captured in national accounts

(Balvanera and Prabhu 2004; Millennium Ecosystem Assessment 2003). Ecosystems and the services they deliver may have economic value because people derive utility from their actual or potential use, either directly or indirectly (Balmford and others 2002; Costanza and others 1997). Despite its importance to a country's natural wealth, the value of ecosystem services is not captured in national accounts because such services are usually considered public goods (box 1.1).

The world's poor depend disproportionately on ecosystem services, and are highly vulnerable to their disruption (DFID and others 2002). With few alternative income sources, their survival and livelihoods are based on small-scale agriculture, grazing, harvesting, and hunting or fishing. Without adequate infrastructure to provide safe drinking water, electricity and fuel, and transportation, people rely on fresh air, natural sources of fresh water, soil, and biodiversity to meet their basic needs.

About 1.3 billion people live on marginal lands (UNDP 2003a); these people are both the victims and the agents of environmental damage (Duraiappah 1998). Yet the role of sustainable resource management and environmental protection for poverty reduction is understated in most countries' poverty reduction and development aid strategies (Bojö and others 2004).

Links to other Millennium Development Goals

Environmental sustainability is inextricably linked to the other Millennium Development Goals (table 1.1). Thus, environmental degradation can have serious repercussions for achieving other Goals and targets. Advances toward one may accelerate progress toward another. Conversely, lack of progress on one may impede movement toward another. In other cases, progress toward one may hinder another.

For example, the pollution and overappropriation of surface water and groundwater resulting from intensified agricultural production can prevent access to clean drinking water, thereby harming human health. Similarly, interventions to meet poverty reduction and food security targets might adversely affect the environment, and hence achieving target 9 (figure 1.1). In still other cases, synergies may be realized where a single intervention or package of interventions creates movement toward several goals more cost-effectively than if applied in a single sector. Clearly, any intervention in one sector must be evaluated for potential synergies or effects, both positive and negative, on another. Of course, tradeoffs and compromises will be required; at the same time, concerted efforts must be made to mitigate the adverse effects of advancing toward one goal at the expense of another.

To illustrate the extent of the environment's importance in achieving the other Goals, the task force devotes the remainder of this chapter to examining in depth the links between the environment and poverty, food security, and health.

Box 1.1

Accounting for environmental degradation and resource depletion

The environment supplies goods and services that support life and human economic activity. Yet traditional national accounts fail to include measures of resource depletion and the costs of environmental degradation. The stock of natural capital includes renewable and nonrenewable resources—forests, mineral deposits, soil nutrients, and energy resources; these constitute a major part of the total wealth of all countries and are particularly important components in many developing economies. When countries "spend" this natural capital by extracting and selling it, they effectively convert natural assets into financial ones—a net zero change in wealth. Standard accounting measures fail to capture the effects of resource depletion, instead counting revenues from the sale of natural resources as earned income.

Accounting for natural resource depletion is conceptually and empirically difficult, but estimates of savings rates can be adjusted to include basic indicators, including energy and mineral depletion and deforestation. As there is no readily accepted, practical method for correcting measured national saving rates for resource depletion effects, this task force presents the World Bank's preliminary attempt at adjusting savings rates for environmental degradation (World Bank 2004a), augmented by an estimate of the economic cost of soil degradation. The results suggest that gross national savings rates greatly overstate increases in the total stock of capital (see table).

Adjusted savings rates, by developing region

Percent of gross national income, 1980–2001

Region	Gross national savings	Adjusted gross savings
East Asia and the Pacific	35	29
Latin America and the Caribbean	19	16
Middle East and North Africa	24	9
South Asia	20	19
Tropical Sub-Saharan Africa	11	3 (1[a])

Note: Adjusted net savings equals net national savings plus education expenditure minus net deforestation and energy and mineral depletion.

a. Nutrient depletion indicators and fertilizer prices were used to calculate soil depletion at around 2 percent of GDP, which would reduce adjusted gross savings to 1.5 percent and adjusted net savings to 1 percent.

Source: Sachs and others 2004.

In the World Bank's corrected savings measures (Hamilton and Clemens 1999; World Bank 2004a), the savings rate taken from the national accounts is augmented by expenditure on education (counted as consumption in the national accounts, but should be counted as investment in human capital), and then reduced according to estimates of the economic costs of deforestation and energy and mineral depletion. Sachs and others (2004) then make an additional correction for tropical Sub-Saharan Africa by measuring the estimated annual loss of three soil macronutrients—nitrogen, phosphorus, and potassium—currently being depleted with each harvest.

The results are dramatic. Whereas Africa's measured national savings rate is about 11.1 percent of GNI, the savings rate net of resource depletion might be only about 1.0 percent of GNI. Whatever the precise number, the conclusion is that Africa and other developing regions experience significantly reduced savings rates as a result of environmental degradation.

Table 1.1

Key links between environmental sustainability and other Goals

Source: DFID and others 2002; UNDP 2002.

Millennium Development Goal	Examples of links to the environment
1. Eradicate extreme poverty and hunger	• Livelihood strategies and food security of the poor often depend directly on functioning ecosystems and the diversity of goods and ecological services they provide. • Insecure rights of the poor to environmental resources, as well as inadequate access to environmental information, markets, and decisionmaking, limit their capacity to protect the environment and improve their livelihoods and well-being.
2. Achieve universal primary education	• Time children, especially girls, spend collecting water and fuelwood can reduce study time.
3. Promote gender equality and empower women	• Time women spend collecting water and fuelwood reduces their opportunity for income-generating activities. • Women's often unequal rights and insecure access to land and other natural resources limit opportunities for accessing other productive assets.
4. Reduce child mortality	• Water and sanitation-related diseases and acute respiratory infections, primarily caused by indoor air pollution, are leading causes of mortality in children under age five.
5. Improve maternal health	• Indoor air pollution and carrying heavy loads during late stages of pregnancy put women's health at risk before childbirth.
6. Combat major diseases	• Environmental risk factors account for up to one-fifth of the total burden of disease in developing countries. • Preventive environmental health measures are as important, and at times more cost-effective, than health treatment.
8. Develop a global partnership for development	• Since rich countries consume far more environmental resources and produce more waste than poor countries, many environmental problems (such as climate change, loss of species diversity, and management of global fisheries) must be solved through a global partnership of developed and developing countries.

Environment and poverty

While consumption patterns of the rich drive overexploitation of natural resources, poor families, in their daily struggle for survival, often lack the resources required to avoid degrading their local environment. Their fragile resources, often poorly defined property rights, and limited access to credit and insurance markets prevent the poor from investing in sustainable environmental management. With few alternative sources of income, they rely extensively on natural resources and ecosystem services to supply such basic human needs as food, fuel, and drinking water. However, overextraction of resources disrupts the environment, causing many to lose access to the ecosystem services on which their survival depends. And when countries draw down natural capital without compensating increases in human, social, financial, or

Figure 1.1
Potential environmental consequences of food security and poverty alleviation strategies

Source: M. Falkenmark, personal communication, 2003.

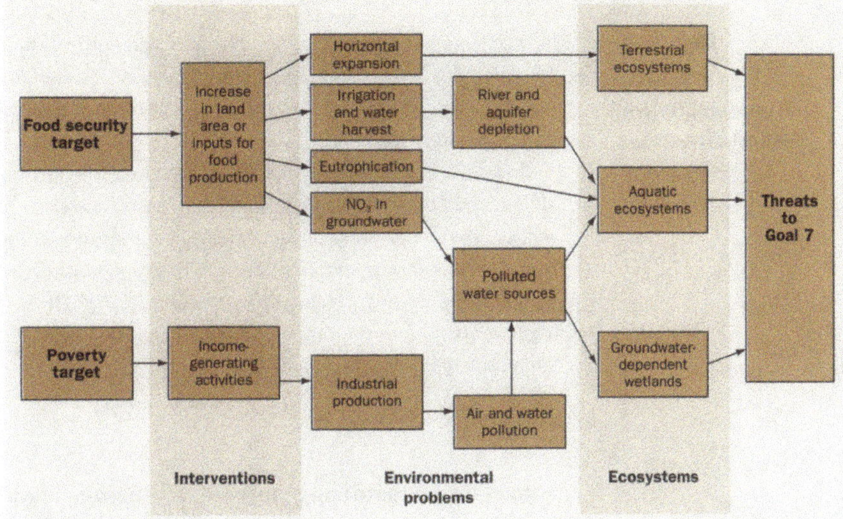

physical capital, they become poorer. Environmental degradation and resource consumption substantially lower national savings rates if incorporated into national accounts (see box 1.1).

Environmental sustainability must be viewed not only as an issue for the poor. In fact, considerable evidence suggests that the greatest threats to environmental sustainability derive from actions taken in the rich countries of the world. The Johannesburg Declaration on Sustainable Development and Plan of Implementation notes that "fundamental changes in the way societies produce and consume are indispensable for achieving global sustainable development" and calls on developed countries to take the lead in "changing unsustainable consumption and production patterns."[3]

Deforestation, for example, is only partly caused by local demand for agricultural land or construction materials. It is even more fundamentally driven by the industrialized world's demand for timber and growing international trade in forest products. Fisheries, mineral deposits, energy supplies, and biodiversity resources are harvested in developed and developing countries alike; however, the preferences and demands of the world's richest countries largely determine the scale and intensity of resource exploitation. Increasingly, governments recognize the often disastrous human consequences of wanton resource exploitation (box 1.2).

Similarly, greenhouse gas emissions in the world's developed countries have largely driven global climate change, which threatens human well-being, ecosystems, and biodiversity. Although developed countries represent only 20 percent of the world's population, they have generated 80 percent of historical greenhouse gas emissions (Watson 2004). If developed countries do not reduce emissions and economic growth in developing countries generates equivalent per capita quantities of greenhouse gases, climate change will

Box 1.2
Shared responsibility for natural disasters: Philippines example

Source: United Nations Statistics Division 2004.

In the Philippines, forest cover shrank 3 percent between 1990 and 2000, from 22.4 to 19.4 percent of land area. As in other countries, loss of forest cover, fueled by widespread legal and illegal logging, has impaired the natural functions of safeguarding watersheds and regulating river flows.

In late 2004, a series of typhoons battered the island of Luzon, triggering severe landslides and floods that killed more than 1,000 people. Deforestation, which had sharply eroded the soil's absorptive capacity, contributed to the scale of the disaster. It was reported that the Government of the Philippines blamed illegal logging for the disastrous scale of the damage (*New York Times*, Dec. 5, 2004); however, both legal and illegal logging practices bear responsibility for the environmental changes that have increased vulnerability to natural disasters.

accelerate, jeopardizing human well-being, biodiversity, and environmental sustainability.

Reducing poverty and achieving environmental sustainability, then, require charting a new path for development between the extremes of resource degradation on the one hand and unsustainable production and consumption on the other. Doing so will require a clear, ambitious set of objectives and strategies and creative, forward-thinking leadership in each nation. The rest of this report provides a foundation for achieving environmental sustainability, and with it the rest of the Millennium Development Goals.

Environment and food security

Food security is integrally linked to environmental sustainability, as all food ultimately derives from ecosystem services. More than 2 billion poor people rely directly on agriculture for subsistence and commercial food production. The ecosystem services critical for production include provision of freshwater for crop irrigation; maintenance of soil fertility through nutrient cycling; provision of crop genetic diversity, crop pollinators, pest control, and climate regulation; and provision of wild collected foods (such as bush meat, fruits, and fish). In West Africa, bush meat is the main source of animal protein; while coastal, pond, and lake fisheries provide the main protein source for people in Southeast Asia and other regions (Millennium Ecosystem Assessment 2004a, ch. 8). Currently, 30 million low-income people—a doubling over the past 30 years—earn their livelihoods primarily from fishing. Consequently, land degradation and depletion of fisheries seriously affect food security (Burke and others 2000; Roberts and others 2002).

Environmental degradation and biodiversity loss are urgent, fundamental problems that threaten the achievement of Goal 1. Where people lack access to modern agricultural technologies, the condition of the local ecosystem determines agricultural productivity and food supply. Thus, sustainable management of terrestrial and marine ecosystems is a prerequisite to global food

In developing countries, major environmental risks account for 18 percent of the disease burden

security. Over the past century, 75 percent of crop genetic diversity has been lost, leaving crops and varieties vulnerable to emerging and spreading disease, pests, and changing environmental conditions (especially those related to climate change). Land degradation continues as a result of inappropriate, intensive agricultural techniques and land conversion related to agricultural extensification. Inappropriate intensification causes salinization of irrigated areas, nutrient and pesticide leaching, and pesticide resistance, while extensification destroys natural vegetation cover and leads to soil erosion and loss of soil fertility, increased withdrawals of groundwater and surface water, and increased agrochemical load (Wood, Sebastian, and Scherr 2000). Unsustainable fish catch and pollution of coastal and marine ecosystems have caused precipitous declines in fish stocks (Pauly and others 2002). These forms of environmental degradation decrease food availability, sometimes irreversibly, complicating efforts to fight hunger.

Environment and health

Environmental degradation adversely affects human health through exposure to bacteria, parasites, and disease vectors (such as mosquitoes and snails); chemical agents (such as heavy metals, particulates, or pesticides in water, food, air, and soil); and physical and safety hazards (such as fire, radiation, and natural disasters) (Bojö and others 2001). However, the world's people are not equally affected. In developing countries, major environmental risks account for 18 percent of the disease burden, about twice the proportion in industrialized countries (Lvovsky 2001). The most vulnerable region is Sub-Saharan Africa, where fully 27 percent of the disease burden is attributed to environmental risks.

Pollution and contamination of air and water are major sources of human illness. Diarrhea, strongly linked to unsafe water and inadequate sanitation, is the leading killer of children under five. At present, 2.2 million people die each year from water contaminated by human feces (UN Millennium Project 2005a; WHO and UNICEF 2000). In most developing countries 90–95 percent of all sewage and 70 percent of industrial waste are dumped untreated into surface water (UNFPA 2001). Heavy metals, such as mercury, have accumulated in many of the world's fisheries, rendering fish stocks unsafe for human consumption. Persistent organic pollutants accumulate in the fats and tissues of animals in the food chain, posing further risks to human health.

Acute and chronic respiratory infections are related to ambient air conditions influenced by the incidence of wildfires, vehicle pollution, and industrial discharge. Indoor air pollution from the use of biomass fuels in poorly ventilated houses has been linked to 1.6 million deaths worldwide (Warwick and Doig 2004).

Many of today's emerging or resurgent diseases, such as malaria, dengue, and mosquito-borne encephalitis, are on the rise because of human disruption of natural ecosystems. Environmental change is cited as one of the six major factors leading to the emergence or resurgence of many of these diseases (Patz and Wolfe 2002; Cohen 2000). Forest clearance, drainage of wetlands, international movement of livestock, illegal wildlife trade, human encroachment on wild areas, human travel, and climatic events can facilitate disease transmission (Daszak, Cunningham, and Hyatt 2000; Patz and others 2004; Harvell and others 1999; Millennium Ecosystem Assessment 2004a, ch. 14) (box 1.3). Insofar as these activities involve environmentally unsustainable practices, better land-conservation measures can reduce the global burden of infectious disease, thereby reducing childhood mortality, malaria, and other diseases and directly benefiting Goals 4 and 6.

There are tradeoffs between the risk of infectious disease and development projects that damage habitat or ecosystems that may be preventing epidemics of infectious disease. To the extent that the risk mechanisms are understood, the potential for preventing or reducing the risk of certain diseases can be achieved through incorporating sustainable conservation measures into

Box 1.3

Protecting natural habitats can alter the risk of infectious disease

Source: Millennium Ecosystem Assessment 2004a.

A range of biological mechanisms are responsible for altering the incidence of infectious disease. Key examples of this disease-ecosystem relationship are described here:

- Dams and irrigation canals provide ideal habitat for snails, the intermediate reservoir host species for schistosomiasis; irrigated rice fields enlarge the extent of mosquito breeding surface, increasing transmission of mosquito-borne malaria, lymphatic filariasis, Japanese encephalitis, and Rift Valley fever.
- Deforestation increases the risk of malaria in Africa and South America, while its effect in Southeast Asia is uncertain.
- Natural systems with preserved structure and characteristics are not receptive to introduction of invasive human and animal pathogens brought by human migration and settlement (this is the case for cholera, kala-azar, and schistosomiasis).
- Uncontrolled urbanization in the forest ecosystem is associated with mosquito-borne viruses (arboviruses) in the Amazon and lymphatic filariasis in Africa. Tropical urban areas with poor shelter and lacking closed-pipe sanitation promote transmission of dengue fever.
- Habitat fragmentation, with subsequent biodiversity loss, increases the prevalence in ticks of the bacteria that causes Lyme disease in North America.
- Zoonotic pathogens (defined by their natural life cycle in animals) are a significant cause of both historical infectious diseases (such as HIV and tuberculosis) and newly emerging ones (such as SARS, West Nile virus, and Hendra virus); in addition, zoonotic pathogens can result in high fatality rates and are difficult to vaccinate against since the primary reservoir hosts are nonhuman (for example, Nipah virus).
- Intensive livestock agriculture, treated with subtherapeutic doses of antibiotics, can lead to the emergence of antibiotic-resistant strains of *Salmonella, Campylobacter,* and *E. coli* bacteria; overcrowded and mixed livestock practices and bushmeat trade have facilitated interspecies host transfer of disease agents, leading to dangerous novel pathogens, such as SARS and new strains of avian influenza.

development plans; therein lies the potential to achieve a double benefit of sustainable human health and environment.

Finally, the world's stock of biodiversity and genetic resources has proven critical in the fight against illness and disease; however, anthropogenic environmental change has threatened these priceless and irretrievable resources. In most developing countries, up to 80 percent of medicines are derived from medicinal plants (Balick, Elisabetsky, and Laird 1996). The growing body of literature on the ethnopharmacopeia of certain regions or societies is further evidence of the importance of biodiversity to human health.

Other links

To reiterate, environmental sustainability is linked in important ways to all of the other Millennium Development Goals (see table 1.1). As the context for all human behavior, the environment influences educational opportunity, gender disparity, water quality and sanitation, as well as the interests of urban slum dwellers and small island and landlocked developing states. Chapter 2 discusses the state of six key elements of the environment that are critical for achieving the Goals.

Chapter 2

Driving environmental change

We live in an era of unprecedented environmental change (Brackett and others 2004). Rapid population growth, economic development, and international economic integration have intensified resource use; in every region of the world, human actions have directly or indirectly increased pressure on the natural environment. Rapidly diminishing forests and coral reefs, increased consumption of scarce water and energy resources, desertification, the spread of invasive alien species, the breakneck rate of biodiversity loss, and the rising threat of global climate change highlight the urgent need to address the complex dimensions of environmental change.

Left unmanaged, the progressive anthropogenic deterioration of environmental conditions will stymie efforts to achieve the Millennium Development Goals. Addressing these problems requires not only scientific evidence of environmental change, but a deeper understanding of the factors directly and indirectly driving this change.

Mitigating and reversing environmental change will require an understanding of its drivers; the agents capable of influencing them; the design of appropriate interventions; and incentives for individuals, organizations, and governments to implement these actions. This chapter moves toward this goal by providing a baseline assessment of environmental change; it identifies key direct and indirect drivers of change in six key elements, or thematic areas, of the world's environment and describes the major challenges that confront each world region. While these six areas are not an exhaustive categorization, they capture many of the most significant resources and processes upon which human health and economic well-being, biodiversity in all its forms, and environmental sustainability ultimately depend. These six areas also provide the framework for this task force's recommendations for improving environmental management (see chapter 4).

Human activities influence the environment through direct and indirect drivers

Human activities, undoubtedly the principal cause of most environmental change, influence the environment through both direct and indirect drivers. Following the Millennium Ecosystem Assessment, this task force broadly defines drivers to include "any natural or human-induced factor that directly or indirectly causes a change in an ecosystem." Direct drivers are essentially the proximate causes of changes in ecosystems; they produce unequivocal and measurable environmental effects. Other drivers are indirect in that they operate diffusely, often by altering one or more direct drivers (Millennium Ecosystem Assessment 2004b, ch. 7).

Without understanding the factors driving environmental change, it is difficult to design effective strategies for environmental management. Conversely, a robust understanding of the drivers behind environmental degradation enables decisionmakers to make informed economic, political, social, and behavioral changes. Furthermore, these drivers operate at a variety of temporal scales; understanding their dynamic relationships can improve coordination and sequencing of interventions.

This chapter focuses on five of the most significant drivers, each of which poses specific challenges to environmental sustainability and requires targeted responses. In addition, natural climatic events cause environmental change at various temporal and spatial scales. The natural disasters that cause most deaths, by world region, are floods (Americas and Africa), drought and famine (Asia), earthquakes (Europe), and avalanches, landslides, and wind-related storms (Oceania) (Millennium Ecosystem Assessment 2004a, ch. 6). Vulnerability to natural disasters is an important factor in designing development strategies and assistance programs (Dilley and others 2005). Certain regions and economies are more vulnerable to climate-related disasters. For example, human vulnerability to drought is high in many African nations and in parts of Asia, where disasters caused by hydrological events are more common (map 2.1). Economic vulnerability to these events, however, is more widespread (map 2.2).

Direct drivers of environmental change

Among the direct drivers of environmental change, five are particularly important.

Land-cover change

Logging, urbanization, conversion to agriculture, road construction, human habitation, and other land-cover changes have emerged as a major driver of environmental degradation. Land-cover change can cause biodiversity loss and impair the delivery of vital ecosystem services, including the water-retaining and flood-attenuating capacity of soil. Changes in land use can severely affect riparian and marine ecosystems, choking coral reefs, rivers, and estuaries with eroded soil and its chemical contents.

Map 2.1

Mortality risk based on vulnerability to drought and hydrological events

Top three deciles of risk

Source: Dilley and others 2005. Reprinted with permission of the World Bank.

Map 2.2
Economic loss risks based on vulnerability to drought and hydrological events
Top three deciles of risk

Source: Dilley and others 2005. Reprinted with permission of the World Bank.

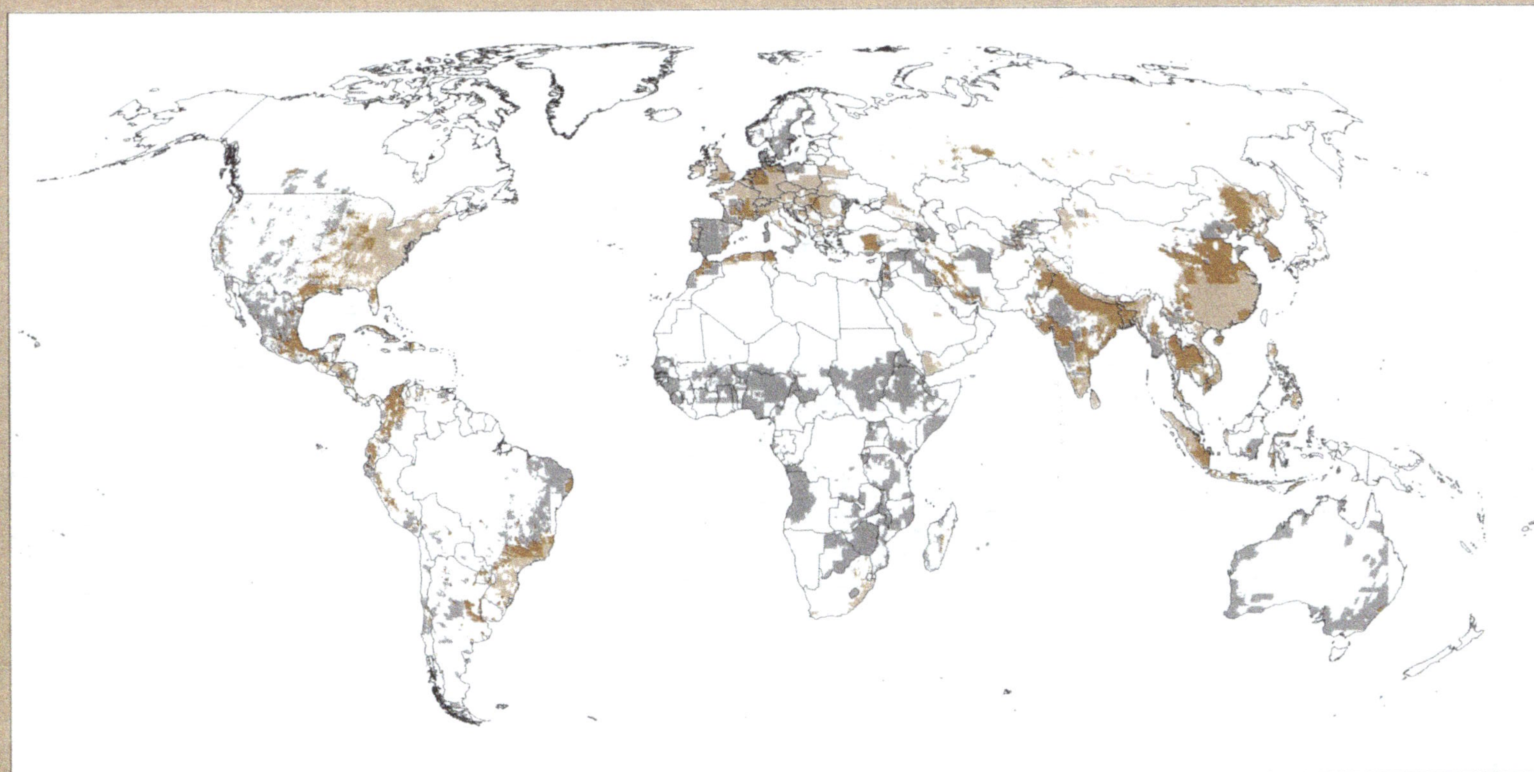

Indirect drivers usually influence one or more direct drivers

Overappropriation or inappropriate exploitation of natural resources

Overuse of limited natural resources can reduce the stocks of even renewable resources below sustainable levels. Overfishing is a clear example of overappropriation, as are overconsumption of water and unsustainable harvesting of timber and other forest products. Inappropriate exploitation includes harvesting techniques that cause extensive collateral damage, such as blast or cyanide fishing or intensive agriculture.

Invasive alien species

Introduced accidentally or purposefully, invasive alien species—non-native species that establish themselves in new environments and then proliferate and spread—can choke out native species, clog waterways and infrastructure, and threaten human health in ways that significantly affect economic performance, poverty alleviation, food security, and biodiversity conservation.

Pollution

Pollution impairs human health, reduces agricultural production, damages ecosystems, and adversely affects nearly every dimension of environmental sustainability. While pollution may result from direct inputs from chemical and other industries, it can also involve direct inputs from such human activities as use of "dirty" fuels and indirect inputs such as use of chemical fertilizers and pesticides and poor waste management.

Climate change

Over the coming years, the single greatest driver of environmental change may be climate change. This emerging threat can disrupt the natural balance of ecosystems and adversely affect development concerns. Small island developing states and the highest latitudes and altitudes have already been differentially affected by the warming trend of the past 30 years and its associated ice melting and sea level rise.

Indirect drivers of environmental change

Indirect drivers of environmental change usually influence one or more direct drivers (Millennium Ecosystem Assessment 2004b, ch. 7). Indirect drivers explain, for example, why land-use patterns are changing, how intensification of trade and transport has fueled species movement, how patterns of production and resource use have shifted, and why these changes might cause more pollution and contribute to global climate change. Because all socioeconomic behavior is embedded within the surrounding natural environments, it is impossible to capture exhaustively the indirect relationships between human activity and environmental change. Drawing on the work of the Millennium Ecosystem Assessment, the task force focuses on the most important indirect drivers.

Demographic shifts have key implications for resource use

Demographic change

Demographic trends with key implications for the environment include population growth, rural-to-urban migration, and shifts in household economic status. The UN Population Division forecasts a population increase (medium variant projection) of 2.6 billion people between 2003 and 2050, yielding a global population of 8.9 billion people, 86 percent of whom will live in developing countries (UNDESA Population Division 2004).

Fertility is highest in the poorest countries and among the poorest people in poor and middle-income societies. Not surprisingly, these countries have the highest levels of unmet needs for family planning and reproductive health services; in concert with other health, education, and gender equality issues, these needs must be addressed through policies and programs that slow population growth and realize synergistic improvements. At the same time, many developing countries are experiencing significant rural-to-urban migration. By 2030, 60 percent of the world's population is expected to live in urban areas (table 2.1). Urban populations are increasing by 0.86 percent per year as rural populations post a similar decline (UNDESA Population Division 2004).

These demographic shifts have key implications for resource use. Population growth increases the demand for such essential goods and services as food, shelter, and energy. It also drives the production of greenhouse gases and airborne particulates, which cause human health and environmental problems. Increased food demand can encourage agricultural intensification, which can reduce pressure to convert natural habitats. Applied inappropriately, however, it can encourage agricultural extensification, pushing settlements into previously undeveloped areas and driving land-use changes that further threaten forests, biodiversity, and soil quality. Population expansion drives overfishing and water-use changes that threaten fisheries and marine ecosystems.

As populations expand in countries that still possess reservoirs of biodiversity, human pressures on ecosystems and the environment will grow. Indeed, population growth around biodiversity zones often exceeds national aggregate growth rates (Cincotta and Engelman 2000). Rural-to-urban migration is causing rapid expansion of cities, with potentially adverse environmental

Table 2.1

Rural/urban population shares in 2003 and 2030 for the six major regions of the world

Percent

Region	Rural/urban population, 2003	Predicted rural/urban population, 2030	Increase in urban population from 2003 to 2030
Latin America and the Caribbean	25/75	16/84	12
Asia and the Pacific	63/37	46/54	45
Africa	62/38	47/53	40
Europe	27/73	20/80	10
North America	23/77	16/84	10
Oceania	26/74	25/75	4
Global	52/48	39/61	27

Source: UNDESA Population Division 2004.

Shifts in consumer preferences may yield positive environmental outcomes

consequences. A large portion of urban growth is also occurring in small and medium-size cities, expanding the number of urban areas facing infrastructure challenges and intruding on their surroundings. Increased demand for transport, energy, and water and sanitation services can fuel air and water pollution. Population growth and distribution interact with consumption patterns to produce cumulative effects (UNFPA 2001).

Economic factors

Economic factors drive environmental change in at least six important ways. First, economic growth increases consumption and production, which intensifies resource exploitation, drives changes in land cover and use, and generates waste. Second, and in contrast to the first, rising incomes are sometimes associated with investments in environmental improvement, cleaner technologies, and more robust environmental policies. Third, extreme poverty can be a powerful driver of environmental degradation, as people in desperate poverty understandably care more about meeting their daily needs than ensuring long-term environmental sustainability. Unfortunately, unsustainable exploitation results in the loss of important environmental services, which can reinforce poverty.

A fourth economic driver of environmental change is environmentally damaging subsidies, such as those affecting the fisheries and forestry sectors, which encourage overproduction or overexploitation of resources (Repetto 2004). Other market failures include ownership problems associated with the tragedy of the commons, externalities, and distorting taxes. In addition, failure to account for resource depletion may provide a misleading picture of economic conditions (Sachs and others 2004). Fifth, increased international trade and financial flows shift production and consumption patterns, as well as patterns of resource use. Increased trade flows and transport may facilitate the movement of invasive alien species and pathogens; generate pollution; and, by intensifying the scale of economic activities, encourage overappropriation or inappropriate exploitation of resources. Finally, exogenous shifts in consumer preferences, such as emerging consumer demand for environmentally friendly goods and services, may yield positive environmental outcomes.

Scientific and technological change

Scientific and technological changes are integrally linked to environmental change and exert both positive and negative effects. New technologies can improve environmental outcomes by increasing efficiency of economic activities or enabling more effective pollution abatement. Such innovations as biodegradable packaging or renewable energies can lessen the environmental effect of production and consumption activities. But new technologies that increase the efficiency of renewable-resource extraction, such as factory trawlers in fisheries, without establishing accompanying regulations for sustainable harvest, result in overextraction. Simple technology changes, such as switching the

Sociopolitical factors affect countries' willingness and ability to invest in environmental protection

mesh size or shape of fishing nets or installing composting toilets, can yield positive results. However, new technologies can also introduce new risks, as controversies over nuclear energy and genetically modified organisms suggest. Technology development also drives changes in economic geography and social structures that in turn influence the demand for resources and the way they are extracted. Finally, remote-sensing technology can improve the monitoring of natural resources and pollution and enforcement of regulations so that more sophisticated management measures are taken.

Institutional gaps

Institutions develop shared norms and expectations for behavior through such arrangements as property rights and the rule of law. They can play an important role in addressing market failures by providing information, aligning incentives, and clarifying responsibilities. Conversely, malfunctioning or absent political and regulatory institutions allow overexploitation of fisheries and other natural resources. Weak enforcement regimes fail to deter damaging forms of extraction, such as poaching or illegal logging. Moreover, limited participation of key stakeholders in the planning and management of sustainable resource use reduces the legitimacy and effectiveness of institutions, policy design, and implementation.

Sociopolitical factors

Sociopolitical factors influence patterns of resource use and affect countries' willingness and ability to invest in environmental protection. Differences in culture and social behavior yield varying consumption and production patterns, and social change can produce unpredictable shifts in resource use. Social conflict within countries causes and results from environmental degradation. Environmental degradation can limit resource availability, strain social systems, and intensify latent social tensions.

At the same time, countries in conflict, especially in the limiting case of war, are unlikely to invest in environmental protection or other public goods. In Iraq, for example, physical damage to water and sanitation infrastructure has increased pollution and health risks, caused waterlogging and salinization, and flooded the Tigris River with a mixture of sewage and industrial waste (UNEP 2004). Tensions between countries can complicate efforts to manage critical transboundary resources and ecosystems, such as watersheds or wildlife habitats. Kashmir is a case in point. Domestic political considerations can also complicate efforts to build multilateral frameworks for cooperation on such global issues as climate change.

Controlling the drivers of environmental change

The drivers of environmental change are influenced by decisionmakers' choices at all geographic and political scales (Millennium Ecosystem Assessment 2004b,

Six key elements provide the framework for the task force's recommendations

ch. 7). For example, an individual farmer may control the agricultural production techniques he or she uses to grow crops; his or her decisions directly affect the environment, which is endogenous to these choices. However, this farmer is also affected by exogenous decisions outside his or her control. For example, commodity prices are influenced by market forces, national sector policies, and policies set by governments of foreign countries. As this example illustrates, direct drivers of environmental change are often endogenous to individual actions, while indirect drivers are usually exogenous determinants of individual choices, reflecting political, economic, or social patterns at a much larger scale.

Key elements of environmental change

Understanding how drivers of environmental change influence the world environment is another critical step toward designing appropriate interventions. This task force focuses on six key elements of the environment:

- Agricultural production systems.
- Forests.
- Freshwater resources and ecosystems.
- Fisheries and marine ecosystems.
- Air and water pollution.
- Global climate change.

The condition of these elements affects human health and economic well-being, biodiversity conservation, and environmental sustainability. The sections below summarize the types of services these six areas have traditionally provided and then assess recent trends and key drivers of degradation. These six elements provide the framework for the task force's proposed recommendations on environmental management (see chapter 4).

Agricultural production systems

Sustainable land-based production is directly tied to the health and well-being of the world's population and the livelihoods and survival of the world's poorest people (UN Millennium Project 2005g). Land-based production systems—crop, orchard, plantation, grazeland, and freshwater aquaculture—claim almost one-third of the Earth's terrestrial surface (Millennium Ecosystem Assessment 2004a, ch. 26). Production systems are carved out of natural ecosystems where, formerly, a wide range of services were provided. These include providing critical habitat for the world's biodiversity, including the inordinately large soil biota, the health of which is fundamental to maintaining land productivity. Intact native vegetation also prevents soil erosion and subsequent siltation of rivers and estuaries, helps soils retain moisture, reduces the risk of flooding, and maintains soil fertility through natural nutrient cycling processes.

Land productivity depends on soil fertility and soil water availability. Good soil condition plays a central role in determining the current state of

Land productivity depends on soil fertility and soil water

ecosystems and their future productive capacity (Wood, Sebastian, and Scherr 2000). Perterbations to natural terrestrial ecosystems, such as land-cover clearing, soil tillage, and application of agrochemicals, disrupt natural water and nutrient cycles, often resulting in a degraded land base. In many small-scale cultivated systems, particularly in shifting agriculture, people use the natural dynamics and processes to maintain a productive balance that conserves biodiversity and maintains soil fertility and functionality.

Recent trends. Around the world, inappropriate land-use practices have caused systematic degradation of soil, water, biodiversity, and other resources (table 2.2). A global survey of land degradation found that about 23 percent of all used land had undergone some degree of degradation, including 38 percent of cropland and 21 percent of permanent pasture (Oldeman, Hakkeling, and Sombroek 1991; Oldeman, van Engelen, and Pulles 1991; Wood, Sebastian, and Scherr 2000). Of degraded lands, 38 percent were lightly degraded, 46 percent moderately degraded, and 16 percent extremely degraded and no longer suitable for agricultural use.

Patterns of land degradation vary by terrestrial ecosystem and across world regions; the types and extent of soil degradation have pronounced regional differences. Between 1980 and 1995, the land area under agricultural production increased 10.3 percent in Africa, 19.8 percent in South America, and 17.1 percent in Asia (Barbier and Burgess 2001). In South and Southeast Asia,

Table 2.2
Types, current extent, and causes of land degradation

Type	Current extent of damage (millons of hectares)	Causes and recent losses
Overgrazing	680	Forced settlement of nomadic herders and short-term incentives cause overgrazing on natural grasslands and managed pastureland. Recent losses have been most severe in Africa and Asia.
Deforestation	580	Major causes are large-scale logging and clearance for farm and urban use. During 1975–90, more than 220 million hectares of tropical forests were destroyed, mainly for food production.
Agricultural mismanagement	550	Water erosion causes soil losses estimated at 25,000 million tons annually. Of 255 million hectares of irrigated land, about 30 million hectares are severely degraded by accumulation of salt; another 80 million hectares are affected by salinization and waterlogging.
Fuelwood consumption	137	In many developing regions, fuelwood remains the primary energy source; much of it comes from natural forests. About 1.8 billion cubic meters of fuelwood are harvested annually from forests and plantations.
Industry and urbanization	19.5	Urban growth, road construction, mining, and industry are major causal factors; valuable agricultural land is often lost.

Source: UNEP 2003.

Land mismanagement has far-reaching implications for ecosystems and people

74.1 percent of agricultural land has experienced moderate, strong, or extreme deterioration from wind and water erosion, as well as chemical and physical degradation (Wood, Sebastian, and Scherr 2000). In Central Asia, desertification and erosion have adversely affected both agricultural and nonagricultural land. For example, desertification affects 66 percent of Kazakhstan's territory (Kazakhstan government and United Nations Country Team 2002). The problem is particularly severe in drylands, which cover approximately 47 percent of global landmass and are home to 35 percent of the global population, disproportionately in developing countries (Secretariat of the Convention on Biological Diversity 2001). Approximately 70 percent of the world's drylands are degraded or exhibit symptoms of desertification (UNEP 2003). More than 1 billion people, half of them below the poverty line, live in dry areas where 90 percent of the original habitat has been converted to pasture, agriculture, or urban settlements (UNEP 2003).

Land mismanagement has far-reaching implications for ecosystems and many sectors of society. Conversion of native habitat to agriculture and pasture threatens biodiversity, leaves soil vulnerable to erosion and fertility decline, interrupts water and nutrient cycles, and increases the intensity of and susceptibility to crop diseases and insect attacks. Eroded soil that ends up in river systems is responsible for clogging hydroelectric dams. In Sudan, for example, total capacity of the Roseires reservoir, which generates 80 percent of the country's electricity, has fallen 40 percent in 30 years because of siltation of the Blue Nile (UNEP 2003). In many regions of the world, overappropriation of freshwater resources to support agriculture has resulted in soil salinization and acidification. Overuse of chemical and biological inputs to production systems often contaminates soil and water, causing local and regional problems in neighboring ecosystems. One result of land mismanagement is generation of extensive wasteland, which, in turn, requires people to seek other natural areas to convert.

Direct drivers of degradation. There are three primary direct drivers of environmental change related to agricultural production systems.

- *Land-cover changes.* Agricultural extensification converts and ultimately degrades natural habitat and marginal lands. Habitat degradation not only threatens biodiversity; it also disrupts the soil's natural regulatory functions, resulting in soil erosion, reduced water-holding capacity, and nutrient depletion, as well as desertification and other forms of land degradation.
- *Overappropriation or inappropriate exploitation of natural resources.* Inappropriate agricultural practices degrade soil, introduce pollutants, and contribute to salinization or desertification. Insufficient use of fallows, crop residues, and sustainable farming techniques has stripped soil of key nutrients. Misuse of fertilizers and modern agricultural techniques has contributed to chemical degradation. In addition, inappropriate

People in extreme poverty often lack resources to invest in soil fertility

irrigation has produced waterlogging and salinization in some areas and depleted groundwater resources in others.

- *Climate change.* An emerging direct driver, climate change may dramatically alter rainfall patterns, leading to more frequent droughts and flooding. In some areas, people have already begun to adapt their production strategies and practices to changing climate by introducing other crops.

Indirect drivers. The most significant indirect driver of land and soil degradation is demographic change. Population growth drives the need for more food and productive employment; in most developing countries, this means expanded agricultural production. Expanded production can be achieved through intensification (which, if conducted inappropriately, degrades the soil) or extensification (which destroys native habitats and contributes to degradation).

Land conversion is also driven by economic distortions; for example, agricultural export subsidies in industrialized countries may drive farmers to extensify production systems to compensate for falling prices. In addition, economic and agricultural policies at national and district levels can create perverse incentives to engage in environmentally destructive practices.

People living in extreme poverty often lack sufficient resources to invest in maintenance of soil fertility. Poverty can also drive people to strip the land without regard for long-term sustainability, even though environmental degradation ultimately reinforces poverty. Moreover, inadequate tenure arrangements and privatization that prevents nomadic people from following traditional migratory patterns can lead to overgrazing. Finally, technology and information gaps, such as inadequate training in soil conservation techniques, perpetuate damaging, inefficient practices.

Forests

The Earth's 3.4 billion hectares of forests directly contribute to the livelihoods of 90 percent of the world's 1.2 billion people living in extreme poverty—and many others, in both developing and developed countries—by providing food, fuel, shelter, freshwater, fiber, bushmeat, and genetic resources (Scherr, White, and Kaimowitz 2003). Recent estimates suggest that forests provide more than 1.5 billion cubic meters of timber and 1.8 billion cubic meters of fuel (wood or charcoal) each year (Matthews and others 2000). Extraction and processing of an array of forest resources—timber, medicines, fruits, and other nontimber forest products—form the basis of many local economies, and timber trade accounts for high percentages of national economies in resource-rich countries, including Indonesia and Brazil. Forest resources are also an important source of cultural services (providing a sense of place for millions of people who grow up in or near them) and revenue from nonconsumptive activities (primarily from ecotourism).

Forests harbor more than half of the world's known plant and animal species and hold half of the world's terrestrial carbon stocks. They also stabilize

Forests and other natural habitats are being transformed at accelerated rates

natural ecosystems, storing carbon, controlling soil erosion, and regulating movement of water through the ecosystem. Clear-cutting eliminates these services and can exacerbate the consequences of natural disasters.

Recent trends. Despite their global and local importance, forests and other natural habitats are being transformed to other uses at accelerated rates. FAO (2001) suggests that, over the last decade, 14.6 million hectares of natural forests were deforested each year, representing a 4.2 percent loss of global forest cover; 97 percent of that forest loss was in the tropics, with the highest losses in Southeast Asia, West Africa, and Central America (map 2.3). Forests are being transformed for large- and small-scale agricultural expansion, cattle ranching, plantations, and other uses. This decrease has been only partially offset by 5.2 million hectares per year of new forest gain through reforestation, afforestation, and natural expansion (mostly in nontropical regions). On average, estimates suggest that the world has lost 9.4 million hectares of forest—an area about the size of Portugal—each year over the last decade.

While more forests are being planted to provide timber and fuel, plantations typically do not provide ecosystem services comparable to those provided by natural forests, which feature complex structures, high species diversity, and a wide range of natural products (Millennium Ecosystem Assessment 2004a, chs. 9 and 21). The contribution of plantations to carbon sequestration and water cycling may be dramatically lower than that of natural forests. However, plantations provide economic benefits to local communities who depend on forest products for their livelihoods, and they do contribute to carbon sequestration.

Deforestation, however, is not the only threat to forests. Forest ecosystems can appear relatively intact, even as certain perturbations change the system, sometimes irreversibly. Potential disruptions include overhunting, selective logging, acid rain, and infestation by invasive plants, animals, and pathogens. The ecological effects of deforestation and forest degradation can be devastating to surrounding communities and neighboring ecosystems.

Direct drivers of degradation. There are five primary direct drivers of environmental change related to forest systems.

- *Land-cover changes.* Land clearing for agricultural use—including crop plantation, commercial forestry, and ranching—presents the greatest threat to forests. Local and regional effects of forest clearing include desertification, imminent loss of primary potable water sources and hydroelectric infrastructure, biodiversity loss, siltation in the watershed network, and introduction of alien species that affect water quantity and quality. These effects, in turn, lead to the emergence of infectious diseases that threaten human health and the loss of indigenous sources of medicine.
- *Overappropriation or inappropriate exploitation of natural resources.* Three-quarters of the world's forests are owned and administered

Map 2.3
Percentage change in forest cover
Aggregate national averages, 1990–2000

Source: World Bank 2004a.

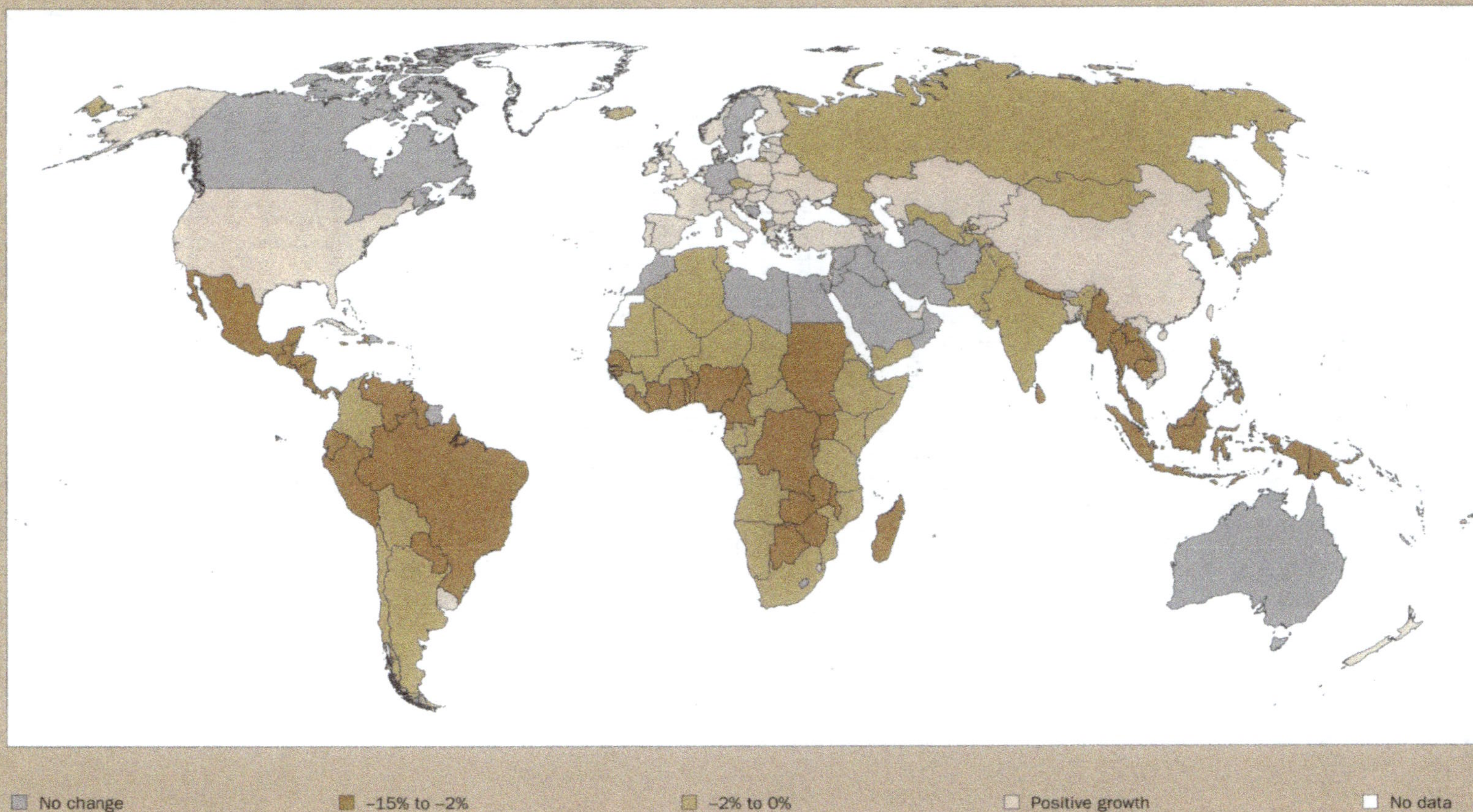

Women have been denied access to decisionmaking and property rights

by governments that have transferred access rights and management authority to large-scale private industry through logging concessions. Few if any incentives for sustainable management of forest resources and ecosystems remain within concession areas. Moreover, the multilateral trading system has no provisions to deter unsustainable, and even illegal, timber extraction.

- *Invasive alien species.* Another direct driver of forest degradation—particularly that of natural and planted forests—are introduced pathogens and insect pests. Urban forests are particularly vulnerable.
- *Pollution.* Acid precipitation, tropospheric ozone, and elevated levels of carbon dioxide threaten forest health by changing the pH of precipitation, altering the behavior of pests, and changing physiological processes.
- *Climate change.* Increased mean annual temperatures, changes in precipitation patterns, and greater frequency of extreme weather events substantially affect the functioning and dynamics of the world's forests and other terrestrial ecosystems. Such changes lead to species loss and redistribution.

Indirect drivers. Major indirect drivers of forest degradation are demographic change, economic factors, and institutional gaps. Population growth has fed demand for land conversion, while growing demand for forest products, tied to international trade and economic growth, has created negative incentives to overharvest. In many developing countries, regulatory and law enforcement regimes have been powerless to stop illegal logging, which accounts for at least half of total timber extraction. Institutional governance has been weak, often excluding forest-dependent communities from both forests and decisionmaking processes. In addition, women, often the primary users and managers of local forest resources, have been denied access to decisionmaking processes and property rights that would allow them to improve the outcomes of resource management policies and projects.

Freshwater resources and ecosystems

Freshwater resources are critical to human survival and environmental sustainability; they provide the fundamental societal functions for human life-support, food production, and energy production, and a transport medium. Freshwater systems support fisheries and other aquatic biodiversity and are essential components of the freshwater cycle, providing important regulating services. By retaining water for periods of time, wetlands, rivers, lakes, and reservoirs help mitigate the danger of flooding. Water uses for ecosystem functions and societal needs are interlinked because they often depend on the same watershed. The integrated management of water resources, which considers the effects of one use upon others, is critical to coordinate supply to the multiplicity of end users.

Over half of the world's natural wetlands have disappeared

Recent trends. Freshwater and the processes that renew it are distributed unevenly across the globe; supply in any given area is strongly affected by ecosystem conditions at a distance of thousands of kilometers. Current per capita availability of water varies considerably with natural and anthropogenic factors; worldwide averages have declined 50 percent over the past 40 years, a trend that calls for significant behavioral changes and new technologies to increase use efficiency (World Water Commission 2000). Since the 1960s, global freshwater use has expanded an average of 25 percent per decade (Millennium Ecosystem Assessment 2004a, ch. 7). By the mid-1990s, water was being withdrawn from rivers, lakes, reservoirs, and wells for household use, industrial supply, and irrigation at a global rate of 3,900 cubic kilometers per year (World Water Commission 2000). Irrigated agriculture accounts for approximately 70 percent of water withdrawals worldwide, consuming 50–100 times more than is used for household needs (Falkenmark 2003; WRI 1998).[1]

About a third of worldwide water use depends on unsustainable water withdrawals, most frequently in Asia, the Middle East, and North Africa. Excessive use of groundwater leads to saline intrusion, ground subsidence, and lowered water tables. In the United Arab Emirates, for example, excessive groundwater pumping has created cones of depression of up to 100 kilometers in diameter, yielding falling groundwater levels, causing wells to dry up, and facilitating the intrusion of saltwater (UNEP 2003). Water resources, both below and above ground, are overappropriated and poorly managed. Humans consume approximately 10 percent of continental runoff, and streamflow is currently overappropriated in 15 percent of the world's land area to below the minimum acceptable level needed for aquatic ecosystem health (generally regarded as 30 percent of a river's flow) (Millennium Ecosystem Assessment 2004a, ch. 7). Surface water has become increasingly polluted, with developing countries suffering the cumulative effect of multiple generations of pollutants, threatening both human health and aquatic ecosystems.

At the same time, degradation of inland water bodies has adversely affected the quantity and quality of available freshwater. Over half of the world's natural wetlands have disappeared.

Direct drivers of degradation. There are four primary direct drivers of environmental change related to freshwater resources and ecosystems.

- *Land-cover change.* Watershed changes that harm freshwater ecosystems include dams for energy generation, channelization and flow diversion for irrigation and flood control, wetland drainage, and groundwater withdrawal. Downstream systems suffer the largest loss of biodiversity since their habitats accumulate the effects from all upstream human activities.
- *Overappropriation or inappropriate exploitation of natural resources.* Excessive water diversion for irrigation and urban use creates shortages

Biological pollution in water is responsible for 2.2 million human deaths annually

and, in some cases, salinization. Currently, approximately 10 percent of water resources are withdrawn from rivers and aquifers, two-thirds of which are consumed during use, primarily in irrigated agriculture. The one-third that returns to the flow is generally loaded with pollutants. Aquatic ecosystems have been considerably affected.

- *Pollution.* Widespread and expanding contamination of surface and groundwater sources by toxic chemicals, organic waste, solid wastes, and suspended solids reduces the safety of natural water. Pollution increases people's dependence on household implements to secure safe drinking water. Biological pollution is responsible for 2.2 million human deaths annually (Millennium Ecosystem Assessment 2004a, ch. 7). Chemical pollution, including naturally occurring arsenic and fluorine, also jeopardizes human health.
- *Climate change.* Natural climate variability, with low and seasonal rainfall in some places, and climatic stochasticity affect surface water availability, increasing floods and drought (see maps 2.1 and 2.2). Climate change may exacerbate these factors, as changes in temperatures are predicted to change the spatial and temporal distribution of precipitation. Climate change may result in increased precipitation at middle and higher latitudes and decreased precipitation in the subtropics, accompanied by more frequent, heavy precipitation events.

Indirect drivers. Major indirect drivers of changes in freshwater resources and ecosystems are demographic change, economic factors, and institutional gaps. Global population growth and rural-to-urban migration increase urban water demand and also push more people into already stressed rural areas. Direct and indirect agricultural subsidies can undervalue water and weaken incentives to manage it sustainably. More broadly, institutional problems related to poorly integrated management of water resources may cause shortages, pollution, and conflict. Moreover, inadequate coordination across sectors and regions results in excessive, inefficiently allocated claims on limited resources. Finally, inadequate infrastructure for freshwater collection, storage, and distribution can create insecurity of supply, generating conflicts of interest.

Fisheries and marine ecosystems

Oceans cover about 70 percent of the planet's surface and are by far the largest habitat for life on Earth. They supply billions of people with food and mineral resources. Deep oceans and coastal waters are the source of most fish catch—a significant protein source for more than two-thirds of the world's population, accounting for 16.5 percent of total animal protein consumed worldwide (Burke and others 2000). Sustainable fisheries play a vital role in achieving the Millennium Development Goals since undernutrition remains the major cause of human mortality, accounting for 30 percent of deaths.

Coastal systems are disproportionately affected by human activity

Marine biodiversity provides critical global ecosystem services: climate control, carbon sequestration, and oxygen generation. A recent study that linked overfishing with climate change showed that sardines play an important role in regulating upwelling ocean ecosystems by devouring large amounts of phytoplankton, which would otherwise cause toxic gas plumes and dead zones upon decay on the ocean floor (Bakun and Weeks 2004). Finally, coastal waters provide cultural and environmental services, frequently supporting tourism and recreation. As the interface between terrestrial and ocean ecosystems, however, coastal systems are disproportionately affected by human activity.

Recent trends. About 40 percent of the world's people live within 60 kilometers of the coastline; as rural-to-urban migration increases, this figure will likely increase dramatically. Fish is an important source of protein for the world's poor, yet 25 percent of all wild fish stocks is underexploited or only moderately exploited (FAO 2003). Overfishing is responsible for 55 percent of marine extinctions (Dulvy, Sadovy, and Reynolds 2003), and accounts for the loss of more than 90 percent of top ocean predators (Myers and Worm 2003). Overexploitation of marine resources, as well as the bycatch of target and nontarget species, can have significant ecological consequences. Fishing down the food chain—catching fish in lower trophic levels—results when large, high-trophic level fish are overexploited, producing an aquatic community dominated by smaller fish (Pauly and others 1998). Coral reefs provide fish and seafood for 1 billion people in Asia alone, yet 80 percent are at risk from coastal development, fishing-related pressures, and climate change (Bryant and others 1998; Roberts and others 2002).

Direct drivers of degradation. There are four primary direct drivers of environmental change related to fisheries and marine ecosystems.

- *Overappropriation or inappropriate exploitation of natural resources.* Destructive and nonselective fishing practices, such as bottom-trawling or dynamite or cyanide fishing, irreversibly damage aquatic ecosystems and adversely affect threatened seabirds, marine mammals, and other marine biota. Overfishing has massively depleted fish stocks and poses a major threat to biodiversity.
- *Pollution.* Organic pollution from ocean dumping and the downstream effects of land-based activities create dead zones and degrade important fish habitat.
- *Invasive alien species.* Marine ecosystems, especially estuaries and bays, are vulnerable to invasive alien species introduced by the exchange of ballast water, which transports some 3,000 animal and plant species each day (UNEP 2003).
- *Climate change.* Climate change threatens to alter the conditions that support the biological diversity of many marine ecosystems. For example,

Air and water pollution threaten marine and inland water resources, soils, and forests

it may increase sea-surface temperatures and thus cause sea levels to rise; decrease sea ice cover; alter salinity, wave climate, and ocean circulation; and lead to bleaching of coral reefs.

Indirect drivers. Major indirect drivers of fisheries and marine ecosystem degradation are demographic change, economic factors, and institutional gaps. Fisheries face increased pressure as population growth and economic development increase demand for marine sea foods. In addition, environmentally damaging subsidies to the fisheries sector stimulate overcapacity and overexploitation. Economic factors and institutional gaps have coincided to produce environmentally damaging outcomes. Ineffective regulatory regimes—poor policies and enforcement, lack of zoning plans, and insufficient attention to scientific advice—have resulted in overexploitation and collapse of key marine populations.

Air and water pollution

Clean air and water are essential preconditions for human life and healthy ecosystems. The range of human activities that emit chemical, biological, and particulate pollutants adversely affect human health. Pollutants can cause brain damage, respiratory illness, cancer, endocrine disorders, and even death. Air and water pollution threaten marine and inland water resources, soils, and forests. They endanger biodiversity by destroying habitat, causing reproductive impairment and generating other population-level effects.

Recent trends. According to the World Health Organization, the six most significant air pollutants are carbon monoxide, lead, nitrogen dioxide, particulates, sulfur dioxide, and ozone (WHO 2000). While criteria air pollutants have decreased significantly over the past several decades in industrialized countries, ambient levels of these pollutants in developing-country cities are much higher, often significantly surpassing WHO guidelines (UNEP 2003).

Indoor air pollution generated by the combustion of biomass and coal for household cooking and heating is a major problem in developing countries, causing chronic and acute respiratory diseases, birth defects, and other sicknesses. An estimated 2.4 billion people burn biomass for cooking and heating, releasing particulates and air pollutants that severely affect human health (Warwick and Doig 2004). Estimates suggest that indoor air pollution kills nearly 2.5 million children each year (WHO 1997); India alone accounts for more than 500,000 of these deaths (Holdren and Smith 2000).

Outdoor air pollution is also a significant problem in developing countries, leading to respiratory illness and morbidity. In Latin America, air pollution causes 2.3 million cases of infantile chronic respiratory sickness each year (UNEP 2003). In Asia, massive quantities of dust, ash, soot, sulfates, nitrates and other particulates have coalesced into a persistent Asian "brown cloud" that hovers over much of the continent, blocking out sunlight and disturbing

Over half of the world's major rivers, lakes, wetlands, and groundwater areas are contaminated

normal weather patterns (UNEP 2003). These forms of pollution drive other destructive processes, such as acid deposition; in southern China, for example, acid rain affects 19 percent of agricultural land (Watson 2004). Persistent organic pollutants, which decay slowly, accumulate in animal fats and pose serious health risks to humans and wildlife alike (UNEP 2003).

Water pollution, driven by household and industrial chemical and biological waste, affects coastal, marine, and freshwater resources. Significant sources of coastal and marine pollution include oil spills and leaks, industrial effluents, sediment, agricultural nutrient run-off, sewage, and other domestic wastes (UNEP 2003). Wastes and effluents from land-based activities are frequently discharged into waters without effective treatment. For example, in Latin America, some 98 percent of domestic wastewater is discharged untreated into the Pacific Ocean (UNEP 2003).

Marine pollution has caused coral bleaching, eutrophication, and bio-accumulation of toxic substances in marine animals. Pollution of freshwater resources is also severe, resulting from untreated sewage, chemical discharge, petroleum leaks and spills, mining residue, and run-off of sediment and nutrients from agricultural fields. Over half of the world's major rivers and associated lakes, wetlands, and groundwater areas are contaminated by pollutants, implying severe problems for human and ecosystem health (UNEP 2003).

Direct drivers of degradation. There are two primary direct drivers of environmental change related to air and water pollution.

- *Land-cover change.* Conversion of natural habitats to agriculture increases soil erosion, causing siltation and the transmission of organic compounds, chemical fertilizers, and pesticides into bodies of water.
- *Overappropriation or inappropriate exploitation of natural resources.* Rising demand for modern energy services—frequently dependent on fossil-fuel combustion—has significantly increased emission of airborne pollutants (sulfur, nitrogen, and carbon monoxide) in many regions of the world. Transportation services—cars, trucks, buses, and motorcycles—are major drivers of urban air pollution, and their contributions are particularly damaging where unleaded fuel is not widely used. Rapidly increasing vehicle fleets in many developing countries have created dangerously unhealthful conditions, especially in large cities.

  Indoor air pollution is driven primarily by household dependence on biomass and solid fuel combustion for cooking and heating; the particulates and toxic gases emitted adversely affect health outcomes.

Indirect drivers. Major indirect drivers of air and water pollution are demographic change, economic factors, institutional gaps, and sociopolitical factors. Rural-to-urban migration has increased demand for energy and transportation services. Economic factors prevent households from investing in improved technologies

In 1999, industrial countries emitted about 10 times more carbon per capita than did developing countries

and infrastructure services. Attempts to improve both air and water quality have been frustrated by weak regulatory and enforcement regimes, which allow industries to pollute waters with toxic chemicals, and political factors that stall development and dissemination of cleaner, more efficient technologies.

Global climate change

A stable climate provides critical regulating and supporting services upon which all ecosystems depend. Climate affects weather patterns and events, agricultural and marine productivity, distribution and population health of species, and energy consumption (heating fuel or cooling technologies). In a complex feedback system, the climate both drives and is driven by the interactions among components of the environment: the atmosphere, ocean, terrestrial and marine biospheres, cryosphere (ice), and soils. Regional weather patterns are strongly influenced by vegetation cover, reflection of solar radiation, air flow, and the water cycle.

Recent trends. Since the advent of the industrial era (around 1750), atmospheric concentrations of greenhouse gases have increased significantly. For example, carbon dioxide has increased from 280 to 370 parts per million (ppm), nitrous oxide from 265 to 312 parts per billion, and methane from 750 parts per billion to 1,750 parts per billion (IPCC 2001; Watson 2004). Sulfate aerosol concentrations, which tend to cool the atmosphere, have increased regionally, stemming primarily from the combustion of coal and oil. In 1999, developing countries emitted about 1.6 gigatons of carbon per year of total fossil fuel emissions, transition economies about 1.7 gigatons of carbon, and industrial countries about 3.1 gigatons of carbon. Industrial countries emitted about 10 times more carbon per capita than did developing countries (figure 2.1). About 1.6 gigatons of carbon was emitted as a direct result of land-use changes, almost exclusively in tropical developing countries (IPCC 2001; Watson 2004).

Figure 2.1
Carbon dioxide emissions per capita, by region, 1970–2015
Metric tons of CO_2

Source: EIA 2004b (fig. 19).

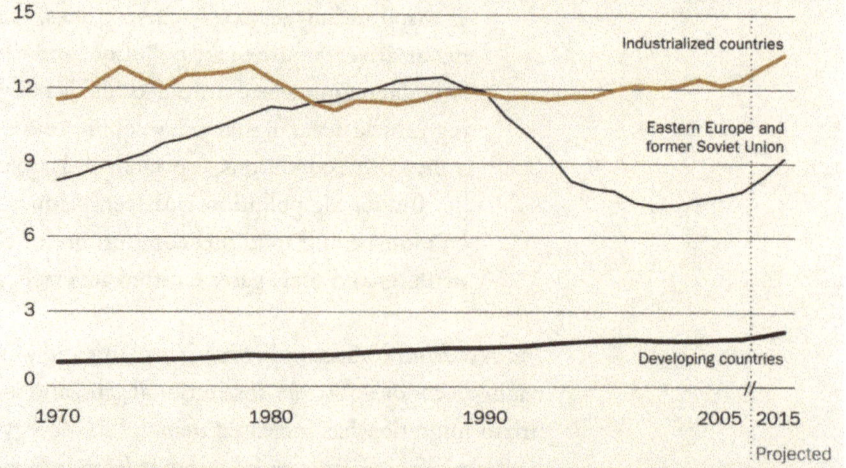

Climate change may exacerbate malnutrition and disease crises around the globe

Historically, more than 80 percent of anthropogenic emissions of greenhouse gases have emanated from industrial countries. Greenhouse gas emissions resulting from energy services and other human activities are projected to increase by up to a factor of 4.5 by 2050 and a factor of 8 by 2100 (table 2.3).

Most scientists now agree that human activities are causing climate change and that, while industrialized countries have caused the problem, developing countries and poor people are the most vulnerable. Over the past century, the Earth's climate has warmed, and though climate change has varied across regions, temperatures have risen on average by 0.6 degrees Celsius. The 1990s was the warmest decade on record. Temporal and spatial patterns of precipitation have changed, and sea levels have risen 10–25 centimeters; most nonpolar glaciers are retreating, and Arctic ice is melting (IPCC 2001; Watson 2004). Between 1990 and 2100, increased atmospheric concentrations of greenhouse gases and aerosols are projected to yield increases in global mean surface of 1.4–5.8 degrees Celsius and a global mean sea-level rise of 8–88 centimeters (IPCC 2001; Watson 2004). Variability of climate patterns and incidence of extreme weather events are projected to increase.

Evidence suggests that low-lying small island developing states and deltaic regions of developing countries in South Asia, the South Pacific, and the Indian Ocean could eventually be submerged, displacing tens of millions of people in the process. Exposure to malaria and dengue fever, already rampant in the tropics and subtropics, could become even more severe, spreading to higher altitudes or northward into the temperate zone. Crop production could significantly decrease in Africa, Latin America, and other developing regions. Freshwater could become scarcer in many areas that already face shortages (IPCC 2001). Climate change will exacerbate biodiversity loss and increase risk of species extinction; one recent study estimated that climate change could lead to the extinction of 15–52 percent of the species analyzed (Thomas and others 2004). Unless mitigated, climate change may undermine efforts to achieve most of the Millennium Development Goals and exacerbate malnutrition and disease crises around the globe.

Table 2.3
Current and projected carbon dioxide and sulfur dioxide emissions, 1990–2100

Year	Carbon dioxide (gigatons of carbon per year)[a]	Sulfur dioxide (metric tons of sulfur per year)	Atmospheric carbon dioxide (parts per million)[b]
1990	6.0	70.9	370
2050	8.5–26.8	29–141	460–580
2100	3.3–36.8	11–93	540–970

a. Assumes no concerted international efforts to protect the climate system.

b. Does not account for climate-induced additional releases of carbon dioxide from the biosphere.

Source: Watson 2004.

Each region faces distinct immediate concerns

Direct drivers of climate change.

- *Land-cover change.* Tropical deforestation eliminates the carbon-fixing capacity of large regions, even as it releases carbon through the combustion of wood and biomass. Subsequent land uses, such as agriculture or livestock production or waste disposal, release methane and nitrous oxides, which further contribute to greenhouse gas concentrations (Watson 2004).
- *Overappropriation or inappropriate exploitation of natural resources.* Increased combustion of fossil fuels (coal, oil, and gas) for electricity, transportation, and heating has released enormous quantities of carbon dioxide and other greenhouse gases into the atmosphere. Manufacturing processes, especially cement production, have also released significant quantities of carbon dioxide and other greenhouse gases as industrial byproducts.

Indirect drivers. The main indirect drivers of climate change are demographic change, economic factors, market failures and distortions, scientific and technological change, institutional gaps, and sociopolitical factors. These forces determine the future demand for energy and changes in land use, which affect emissions of greenhouse gases and aerosol precursors; these, in turn, result in changes in the Earth's climate. Demographic and economic factors, especially population and economic growth, have increased demand for energy and transport services and magnified land-use pressures. Market distortions, such as energy subsidies, have contributed to the problem, as have numerous market failures associated with the systematic underprovision of global public goods. These failures, in turn, are related to institutional gaps and sociopolitical factors that have obstructed efforts to develop effective national and international strategies to deal with related transboundary problems. While much of the science and technology necessary to address climate change already exists, political factors often prevent implementation of the findings. It is a shared responsibility of countries to overcome these obstacles to slowing or reversing the degradation of the earth's atmosphere.

Regional diagnosis of environmental conditions

The preceding discussion provides a global view of key environmental conditions, drivers, and challenges. All regions face common global problems, including climate change, biodiversity loss, and worldwide decline of fisheries; the interconnectedness of the world environment means that people everywhere have a stake in issues anywhere. At the same time, geologic and climatic variations across watersheds and ecosystems mean that each region faces distinct immediate concerns. This section briefly describes the main issues facing developing regions of the world and highlights those for which environmental interventions are needed most. As map 2.4 illustrates, overall progress toward environmental sustainability varies considerably within these regions.

Map 2.4
Environmental sustainability index by country, 2002

Source: World Economic Forum, Yale Center for Environmental Law and Policy, and CIESIN 2002.

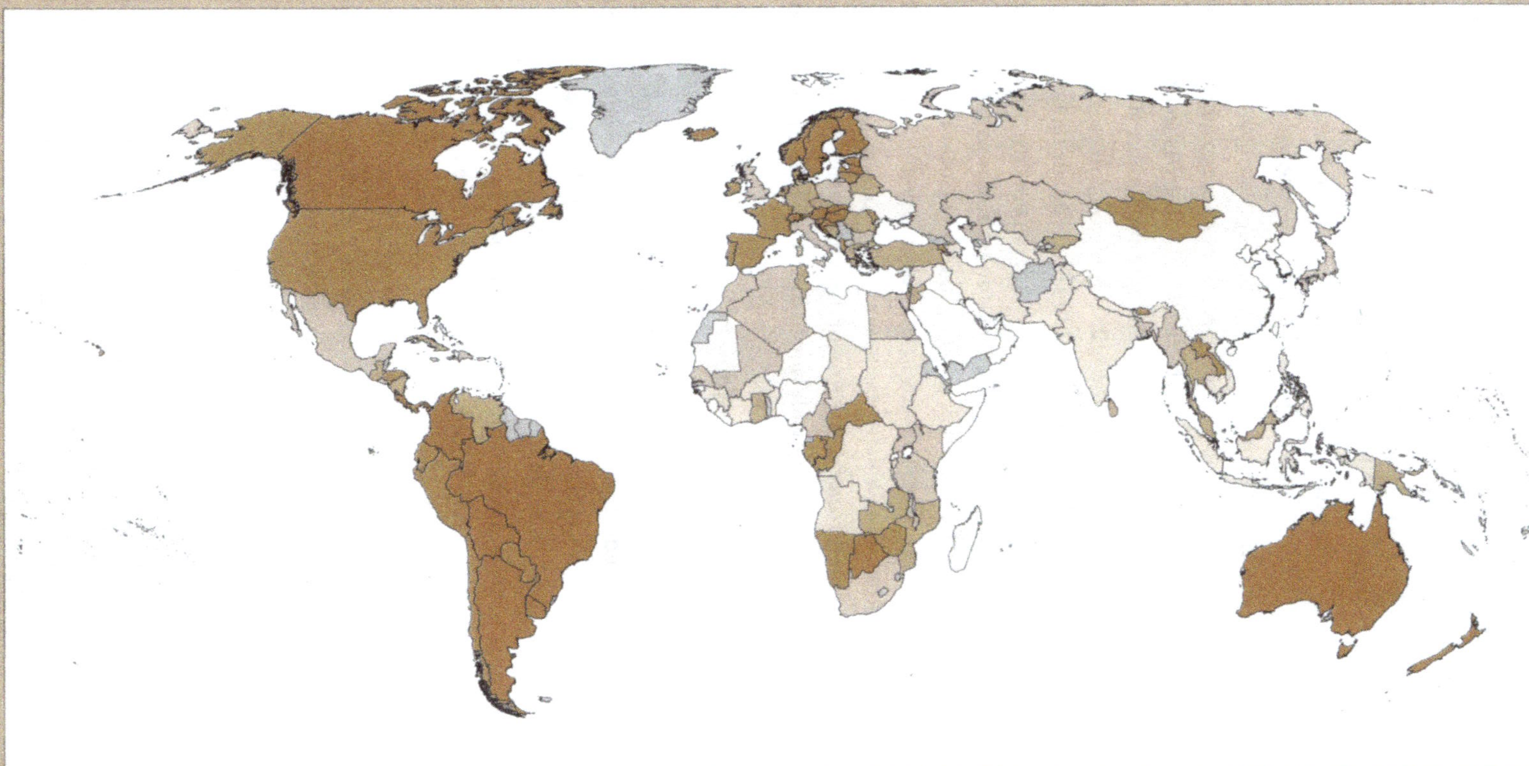

23.9–40.6 | 40.7–45.7 | 45.8–49.4 | 49.5–53.0 | 53.1–57.9 | 58.0–73.9 | No data

Latin America

The most pressing issues facing Latin America are deforestation, pollution, and damage to coastal and marine ecosystems. These trends threaten what is one of the most biologically diverse regions of the world: Together with the Caribbean, Latin America is home to more than 50 percent of the world's flora and fauna (UNEP 2004).

At present, development affects 50 percent of Latin America's land base (UNEP 2004). One *Global Environmental Outlook* scenario projects that as much as 80 percent of the region's natural land cover will be converted by 2032, with potentially disastrous consequences for forest cover and biodiversity conservation (UNEP 2004). Over the past 30 years, Latin America and the Caribbean has accounted for more than 40 percent of global forest decline; between 1990 and 2000, the region lost nearly 47 million hectares, second only to Africa (UNEP 2004). In Brazil, much of this decline was driven by conversion to agricultural land, particularly for soybean cultivation.

Air and water pollution also pose serious threats. More than 80 million people suffer from chronic respiratory sickness related to poor air quality (UNEP 2003). Sixty percent of the region's population lives within 100 kilometers of the coast, which pressures coastal resources, generating many tons of waste that flow directly into the natural environment. Surface and groundwater quality has declined regionwide due to untreated agricultural and domestic wastewater.

Small island developing states, including Caribbean and Pacific islands

Small island developing states are especially vulnerable to environmental change, much of which is driven by factors beyond their immediate control. Key issues include climate change, the health of marine ecosystems, the threat of alien invasive species, and pollution. Global climate change has led to rising sea levels and increased salinization, coral bleaching, more frequent incidence of mosquito-borne diseases, and increased frequency and random patterns of storm systems and coastal erosion. Overfishing, climate change, and mismanagement threaten the health of marine systems. Heavy nautical traffic and exposure to international trade have left many island states vulnerable to invasive alien species, many of which arrive in ballast water exchanges from transport vessels. These species threaten native biodiversity, agricultural production, and water availability. Finally, the sensitivity and scale of small island ecosystems mean that even small amounts of pollution can produce significantly adverse environmental effects.

Africa

On the African continent as a whole, in 1990, at least 13 African countries suffered freshwater stress or scarcity (UNEP 2004). Related issues include siltation of rivers and hydroelectric reservoirs and sharing of surface waters—more than 70 percent—by two or more states, complicating efforts to manage resources sustainably. Improved water management is a direct prerequisite to achieving the Millennium Development Goals. More than 20 percent of the continent's dietary protein derives from freshwater sources; in early 2004, an additional 405 million Africans needed better access to safe drinking water in order to reach target 10 (UNEP 2004). Finally, indoor air quality remains a significant health issue for the region, with traditional fuels accounting for more than half of total energy use (63.5 percent in 1997).

The major environmental issues facing Sub-Saharan Africa are soil and land degradation, depletion of forests and freshwater resources, and poor air quality (Hirji and Ibrekk 2001). Desertification is a growing problem as marginal areas and wetlands are rapidly cleared for agricultural production. Land-use changes and unsustainable farming practices have facilitated wind and water erosion and soil degradation on more than 60 percent of croplands (UNEP 2003). Continentwide, forest declines have resulted from inappropriate economic development strategies and lax implementation of forest protection regulations. Eight of 11 major international watersheds have lost more than 90 percent of forest cover (Hirji and Ibrekk 2001).

Middle East

In the Middle East, the major issues are declining per capita water resources, loss of arable land, pollution-related health problems, and weak environmental institutions and legal frameworks (World Bank 2004b). Overgrazing and forest-cover loss have degraded and exposed soil to erosion by wind and water. Combined with gradual salinization from inappropriate agricultural practices, these changes have contributed to desertification.

This arid region also suffers from chronic water stress and scarcity, exacerbated by unsustainable levels of extraction, contamination, and conflicts over use. Currently, 80 percent of the Middle East population lives in water-stressed areas; annual renewable water resources per capita are expected to drop from 1,045 cubic meters per year in 1997 to 740 cubic meters per year in 2015 (World Bank 2004b). More than 88 percent of water is used for irrigation, and rapid urban growth has placed new demands on increasingly scarce resources (World Bank 2004b). Finally, human health is threatened by declining air quality as a result of rising energy demands, inefficient use of fossil fuels, and industry emissions.

South Asia

South Asia's most serious environmental issues are freshwater pollution and scarcity and soil and land degradation. Several of the region's rivers, including the Ganges, are among the world's most polluted; most rivers in Nepal's urban areas, for example, are too polluted for human use. In Bangladesh, arsenic-contaminated groundwater could be responsible for the deaths of 20,000 people each year (UNEP 2003). Another serious issue is excessive appropriation of water for irrigation, with accompanying decreased annual availability per capita.

Fifty-three percent of land in South Asia is desertified, with reduced agricultural production (UNEP 2001). Agricultural extensification has pushed agricultural activity into increasingly marginal and fragile areas. Overgrazing, overplowing, and vegetation removal have increased the region's vulnerability to desertification. For example, a recent survey found that approximately 53 percent of India's land area has been damaged by wind and water erosion, salinization, and other forms of land degradation (India 2001).

Central Asia

Central Asia's major environmental challenges involve land-cover change and freshwater degradation. Desertification affects more than 66 percent of Kazakhstan's territory (Kazakhstan government and UN Country Team–Kazakhstan 2002). In Tajikistan, erosion affects 30 percent of land area, and desertification claims 1 percent of arable land each year (UNDP 2003c). In Kyrgyzstan, more than 88 percent of agricultural land is degraded, and salinization affects 75 percent of arable land (UNDP 2003b).

Availability and sustainable management of water resources are another major issue. Contamination by organic pollutants, likely resulting from poorly treated industrial and agricultural wastewater, is a serious problem in several countries. In Kazakhstan, for example, more than 200 million cubic meters of polluted wastewater are discharged into surface reservoirs; the government reported that "the natural ability of most watersheds to self-clean . . . seems to have been exhausted" (Kazakhstan government and UN Country Team–Kazakhstan 2002). Another critical issue is management of scarce water resources; for example, volume of the Aral Sea has shrunk by more than 70 percent because of poor irrigation practices and inappropriate regulation of water claims.

East Asia, Southeast Asia, and the Pacific

East Asia's major challenges involve soil and land degradation, deforestation, and poor urban air quality. Overgrazing, overplowing, and removal of native vegetation have accelerated desertification. While some East Asian countries have managed to control deforestation, forest cover in nine Southeast Asian countries declined between 1990 and 2000 (UNESCAP 2003). These trends impede efforts to conserve the region's abundant biodiversity (UNEP 2003).

Coastal and marine areas also face serious challenges. Waste dumping and municipal effluents threaten coastal ecosystems, which bear high pollutant loads and the cumulative effects of tourism and inappropriate fishing practices. Overfishing is a serious problem, exacerbated by such destructive techniques as blast and cyanide fishing, which destroy coral reefs and poison fish habitat.

Across the region, rapid urbanization and industrialization have produced high levels of pollutants, which generate acid rain and damage human health. Scientists consider the Asian "brown cloud" a serious threat to regional climate patterns (UNEP 2003).

Why has environmental sustainability been elusive?

It is a paradox that environmental degradation continues in most regions of the world despite more than three decades of unprecedented international attention. The 1972 Stockholm Conference on the Human Environment focused world attention on the human environment; the 1987 Brundtland Report (WCED 1987) followed, culminating in high-level commitments made at the 1992 Earth Summit in Rio de Janeiro and Agenda 21, which were reaffirmed at the 2002 World Summit on Sustainable Development in Johannesburg, South Africa. Over this period, there has been no shortage of normative and political agreements, high-level pronouncements, and public commitments to reduce environmental degradation (table 3.1).

Of course important progress has been made. For example, the Montreal Protocol has successfully curbed emissions of ozone-depleting substances, many countries have experienced substantial improvement in air and water quality, and large shares of land ecosystems have been placed under protection. Still, the overall picture remains one of enormous concern. Regardless of which indicator one picks for measuring environmental sustainability, most regions are not on track to halt environmental degradation, and some have witnessed dramatic environmental degradation.

Why has the world not progressed farther toward achieving environmental sustainability? Why do most reports on the environment start, like this one, with a gloomy assessment of its current state? Finding compelling answers to these questions is critical for developing a framework of action that can help countries and the world move toward Millennium Development Goal 7.

The direct and indirect drivers of environmental degradation identified in chapter 2 continue unabated, even in the face of mounting evidence of their negative effects on human health and well-being. However, as this task force describes in Part 2, tools and policies exist to attenuate or even mitigate each

Table 3.1
Chronology of major multilateral environmental agreements

Agreement	Year adopted or opened for signature
Convention on Wetlands of International Importance Especially as Waterfowl Habitat (Ramsar)	1971
Convention on International Trade in Endangered Species of Wild Fauna and Flora	1973
United Nations Convention on the Law of the Sea (UNCLOS)	1982
The Vienna Convention for the Protection of the Ozone Layer and its Montreal Protocol on Substances that Deplete the Ozone Layer (Montreal Protocol)	1985, 1987
Basel Convention on the Control of the Transboundary Movements of Hazardous Wastes and Their Disposal (Basel Convention)	1989
Convention on Biological Diversity (CBD) and accompanying Agenda 21, the United Nations Conference on Environment and Development	1992
United Nations Framework Convention on Climate Change (UNFCCC)	1992
United Nations Convention to Combat Desertification (UNCCD)	1994
Rotterdam Convention on the Prior Informed Consent Procedure for Certain Hazardous Chemicals and Pesticides in International Trade (Rotterdam Convention)	1998
Millennium Declaration	2000
International Treaty on Plant Genetic Resources for Food and Agriculture	2001
Stockholm Convention on Persistent Organic Pollutants (Stockholm Convention)	2001
Johannesburg Declaration on Sustainable Development and Plan of Implementation, World Summit on Sustainable Development	2002

driver. Past experience, available science, and promises of improved technologies suggest strongly that environmental damage can be contained without jeopardizing countries' prospects to end poverty. While the reasons for lack of progress differ by country and region, common factors can be identified. In this chapter, the task force focuses on what it considers the most important ones.

Lack of clear operational objectives

As argued in chapter 1, environmental policy is often obscured by a lack of clear operational targets. For example, Poverty Reduction Strategy Papers that mention environmental issues do not tend to set short- and long-term targets (Bojö and others 2004). Thus, environment strategies cannot be monitored; in turn, they become difficult to implement and tend to fall by the wayside in the presence of funding or other constraints.

In cases where targets have been set, such as the Montreal Protocol targets to reduce emissions of ozone-depleting substances, the U.S. clean air and water acts, or European initiatives to curb acid deposition, they have succeeded in focusing the attention of policymakers and galvanizing development of new technologies. For example, several high-income countries have successfully introduced increasingly stringent standards on air quality to reduce emissions

A primary challenge is to set clear priorities and objectives at national and international levels

of sulfur dioxide and nitrous oxides. Often the tightening of standards was announced with long lead times to allow the private sector to adapt by improving technologies to filter air. These changes, in turn, have led to cost savings for private enterprises and the public sector alike.

All too often, however, operational targets are not set since environmentalists fall into the trap of making "better" the enemy of "good": Scientists and policymakers search for a perfect understanding of the challenge and all ramifications of corrective action instead of implementing strategies based on the best available knowledge; as a result, policy changes and investment in environmental management are impeded. Pragmatic approaches that focus on achievable outcome targets are required, while allowing sufficient room for course corrections, based on the principles of adaptive management, as more information becomes available.

In short, lack of clear objectives renders a goal-oriented approach to environmental policy impossible. A primary challenge is to set clear priorities and objectives at both the national and international levels. National targets should be set at a level that responds to global standards, while being commensurate with existing and potential national capacities—economic, technical, social, and institutional. Once operational targets have been set, many sources of knowledge can be drawn on to guide how the targets are to be achieved.

Insufficient direct investment in environmental management

Addressing the direct drivers of environmental degradation and reversing the loss of environmental resources require direct investments in improved environmental management. Chapter 4 describes some of the most important investments and associated policy changes. Examples include improving soil and water management for agriculture, treating wastewater to reduce water pollution, and developing measures to promote reforestation, restore coral reefs, and reduce air pollution. While such measures are well known, scant public and private resources are invested in most developing countries to improve environmental management.

A key challenge involves limited financial resources. In the cases of most low-income and some middle-income countries, domestic resources may be insufficient to cover the full range of needed investments in social services, infrastructure, and improved environmental management. As a result, poor countries are often forced to develop environmental strategies that cannot be implemented. In such cases, sound financing strategies are required that maximize mobilization of private and public domestic resources and draw on external financing as needed (chapter 6).

A related constraint is the incomplete understanding of specific interventions to improve environmental management, such as measures to combat desertification or halt the degradation of mangroves and coral reefs. In contrast to the public health field, for example, where detailed and tested treatment

International organizations working on environmental issues must shift their focus toward country-level advice

protocols are publicly available for every major disease, the environment field has few comparable inventories of best practices. This results, in part, from the complexity of environmental management, which makes it impossible to apply a one-size-fits-all approach. However, better inventorying of successful measures is needed for developing customized investment programs for improved environmental management at local, national, regional, and global scales (chapter 6).

To this end, international organizations working on environmental issues must shift their focus from normative work toward country-level advice on how to prepare and implement strategies for improved environmental management. Such technical support to countries must include advice on the macroeconomic programming of expenditures for environmental management in medium-term expenditure frameworks or other planning frameworks for public investment. Otherwise, it will be difficult to make available the resources required to improve environmental management.

Poor integration into sector policies

Despite the insistence of the 1992 Agenda 21, the 2002 Johannesburg Declaration on Sustainable Development and Plan of Implementation, and most multilateral environmental agreements on the need to integrate the principles of environmental sustainability into sectoral policies, this rarely happens in practice. All too often, national environment strategies—including action plans prepared for implementing major conventions—are developed in isolation from other sectoral strategies that strongly affect environmental sustainability. Lack of political will and difficulty in changing entrenched habits are frequently cited reasons for this quagmire; more proximate causes include:

- Lack of policy space to evaluate tradeoffs between environment outcomes and social or economic objectives.
- Ill-adapted policy formulation and implementation processes.
- Well organized political opposition to change.
- Poor understanding of the social, economic, and health costs of environmental degradation.

While synergies and "win-win" strategies exist in some areas, tradeoffs are common among environmental and social or economic objectives. For example, steps that may appear highly beneficial to a national economy in the short term—increasing fisheries harvests, converting complex forests into mono-specific plantations, or building roads or hydroelectric plants—can have immediate adverse affects downstream or longer-term negative consequences that cannot be easily mitigated. Ironically, these negative effects will, over the long run, damage the very activities they were designed to enhance.

Most countries lack sufficient institutional mechanisms with which to identify and evaluate tradeoffs between development strategies. More often, environmental sustainability is relegated to a sectoral issue assigned to one or

Strategic environmental impact assessments can be an effective tool

more line ministries, with few institutional mechanisms for identifying and resolving potential conflicts between environment and other strategies. For example, most countries lacks forums where senior policymakers and key ministry representatives can weigh the environmental cost of a new infrastructure project against expected economic gains. Likewise, most development strategies are constructed from sectoral strategies, with poor coordination across sectors. While this approach may be acceptable for health interventions to reduce maternal mortality rates, it is inappropriate for crosscutting environment issues. The implication is that mechanisms for developing national poverty reduction strategies or similar development frameworks must be fundamentally revisited, as proposed in chapter 6. In addition, strategic environmental impact assessments can be an effective tool for resolving such tradeoffs (chapter 5).

Finally, opposition to improved environmental management may be better organized than those who stand to benefit from reduced environmental degradation. For example, well organized factory owners may resist tougher environmental standards that would benefit the rest of the country. Likewise, environmentally damaging subsidies for energy consumption, transport, or water use may be difficult to eliminate in the face of strong organized opposition. Overcoming such well organized opposition ranks among the toughest challenges for environmental policymaking and requires tailored responses. Such responses should build on discussing publicly the country's key environmental, social, and economic objectives; improving access to environmental information; and identifying needs and mechanisms for compensatory transfer payments or services to mitigate adverse environmental effects (see chapter 5).

Ultimately, all stakeholders must recognize that the natural environment is the critical infrastructure upon which human health and economic well-being depend. It is within this context that discussions of tradeoffs—which imply concessions by all sides of an issue—should occur.

Inadequate institutional capacity, misalignment of goals, and poor governance

In most developing countries, environmental ministries and protection agencies are severely understaffed and lack sufficient technical expertise and equipment to conduct their work effectively. They tend to lack access to environment experts who can assist in the design of investment strategies that minimize tradeoffs and exploit opportunities for synergies.

This challenge of insufficient technical expertise extends beyond the inadequate numbers of environment experts employed in national and local governments; it also includes their training, which often lacks the breadth and technical rigor required to implement sound environment strategies. Likewise, insufficient numbers of adequately trained extension workers exist to promote farming practices that minimize environmental damage, sustainable fishing practices, and other techniques that can improve environmental management.

Countries must invest in institutional capacity building

Thus, as discussed in chapter 5, countries must invest in institutional capacity building; institutions charged with environmental management must train experts to develop and implement national policies, as well as provide on-the-ground advice to farmers, fishers, foresters, and pastoralists. A particular challenge is strengthening local institutions; the municipal or provincial governments that must implement environmental policies tend to be the most severely underresourced.

Insufficient capacity is compounded by the fact that the institutional goals of many organizations charged with developing environmental policies are poorly aligned with environmental objectives. Institutional goals tend to focus on process targets, such as preparing a new action plan under an international convention or holding a series of workshops, instead of outcome objectives, which require long-term strategies, such as reducing nutrient load flows into a freshwater lake. Despite best efforts many national action plans are not operational and cannot be implemented.

Another closely related issue is poor governance, which manifests in high levels of corruption, poor accountability of policymakers for their decisions, and failure to enforce environmental laws and regulations. Corrupt decisionmakers often condone illegal logging or fishing practices that can have a devastating effect on the local environment. Thus strengthening of institutional capacity must be complemented by improved governance, as outlined in the UN Millennium Project's *Investing in Development* (UN Millennium Project 2005e).

Widespread market failures and distortions

As discussed in chapter 2, market failures can create or exacerbate environmental degradation by misaligning the incentives of individuals or companies with the interests of society at large. This is one reason why private sector expertise and resources are often not marshaled to achieve environmental sustainability. Most of these market failures are well understood and policy responses have been proposed for some. However, available responses can be difficult or even impossible to implement in practice.

To date, no practical approach exists to fully adjust national accounts by accounting for the consumption of nonrenewable natural resources, such as oil or minerals, and the cost of environmental degradation caused (for example, by deforestation or water pollution). Most attempts to implement "green accounting," such as the UN Integrated System of Environmental and Economic Accounting (United Nations and others 2003) or the World Bank's adjusted net savings[1] (Hamilton and Clemens 1999; World Bank 2004a), focus on small subsections of an economy. That no developed or developing country has implemented comprehensive green accounts is not wholly caused by a lack of will or effort. Instead it reflects deep technical and measurement issues that are unlikely to be fully resolved in the near future. Hence partial solutions are required to improve the accounting of environmental degradation

Solutions are required to improve the accounting of environmental degradation

in national accounts. A promising example is the physical flow accounts that have been used successfully in several countries in southern Africa (Lange and others 2003).

Both theory and practice show that sound environmental management requires that people living off an ecosystem have a stake in its preservation through some form of property or tenure rights. Successful examples illustrate that, where communities or individuals were granted such rights, substantial environmental benefits were created. However, the list of failed projects is equally long. In many situations, particularly in rural Africa where land tenure systems are especially complex, no tested strategies exist for addressing this market failure. More research is therefore required to better understand these types of market failures and to propose solutions. As discussed below, this problem becomes particularly difficult to resolve in cases where more than one country is involved in managing an environmental resource.

Other market failures that are difficult to resolve result from situations where some form of transfer payment is required between upstream and downstream residents. For example, people living in river floodplains may wish to compensate farmers living upstream for planting trees to reduce the risk of flooding. Economic theory is clear on the need for effective transfers in such situations, but few practical models are available, particularly in low-income countries where limited institutional capacity places an important constraint on countries' ability to implement complex transfer systems. This issue becomes even more complex in the case of transboundary ecosystems or global environmental change, where international transfer payments may be required.

Finally, market-distorting subsidies may be difficult to remove because of well organized opposition, as discussed above, or because they are designed to reduce poverty or serve well identified social objectives. For example, subsidizing motorized transport, electricity consumption, or water for irrigation may be necessary to support the poor. In practice, however, few subsidies are well targeted and end by benefiting many who do not need them, often at the expense of the poor they intended to help. The principal unmet challenge is to design efficient subsidies that are well targeted toward the groups they are meant to support while minimizing their adverse effect on the environment.

Underinvestment in science and technology

Science and technology are critical for achieving environmental sustainability in at least three ways:

- Improving understanding of the environment to inform the design of policies and investments.
- Monitoring the state of the environment and progress toward achieving specific environmental objectives.
- Developing technologies to mitigate environmental degradation by curbing pollution or improving the efficiency of resource use.

Investments in science and technology fall far short of what is needed for environmental sustainability

In each of these three dimensions, investments fall far short of what is needed to ensure environmental sustainability.

First, major advances have been made in certain areas to harness the best science for improving understanding of the natural environment and informing the policy process. This has been achieved where there are strong national and international scientific research programs and independent, policy-relevant scientific assessments, coupled with effective outreach and communications strategies. In the areas of climate change and biodiversity, well coordinated international research programs include the International Geosphere Biosphere Program, World Climate Research Program, Diversitas, and International Human Dimensions Program; successful independent, international assessments include the International Ozone Assessments, International Panel on Climate Change, and Millennium Ecosystem Assessment.

These initiatives assess and synthesize existing knowledge that is policy relevant, but do not themselves generate new scientific research. They do, however, identify key knowledge gaps and areas for increased research. Unfortunately, environmental scientists too often focus on fundamental research at the expense of applied work of high relevance to policymakers. This important gap needs to be filled. In addition, most decisionmakers, especially those in developing countries, lack effective access to scientific advice on environmental sustainability that would allow them to make informed decisions. To this end, mechanisms need to be set up to provide timely, high-quality science advice to policymakers, as described in chapter 5. Investments should also be targeted at developing indigenous cadres of scientific researchers in developing countries to build local capacity for research, assessment, advice, and policy implementation.

Second, countries' ability to effectively monitor environmental change at the local, national, and regional levels remains inadequate. This is particularly the case for low-income countries, where hydrological and other monitoring systems have deteriorated over the past decades (UNESCO–WWAP 2003). Existing ground-based monitoring systems must be strengthened to improve scientific understanding of the environment, inform the choice of environmental strategies, and track progress toward achieving the objectives. Information from ground-based monitoring systems should be coupled with that obtained from satellite systems. While better monitoring is critical, incomplete information must not serve as an excuse not to act now by investing in improved environmental management.

Third, investments in environmentally friendly technologies are insufficient. With growing per capita incomes and populations, consumption of environmental resources and services may exceed the carrying capacity of ecosystems. Likewise, pollution levels may rise beyond sustainable levels unless technologies are developed to decouple rising consumption levels from pollution and demand for ecosystem services. A good example of potential benefits from improved technologies is the energy sector. Several options exist for minimizing pollution and demand for ecosystem services associated with energy use, such as fuelwood. These include replacing fossil fuels with fuel cells, solar cells, wind power, and geothermal and other energy sources that can dramatically reduce greenhouse gas emissions, air pollution, and the need for fuelwood. Currently, such renewable energy sources are not commercially viable, in most situations, and require ongoing research to improve efficiencies and lower production costs. Despite the enormous benefits that can derive from improved energy technologies, public and private funding for research on renewable energy sources has fallen over the past decade. Other areas where environmentally friendly technologies are not matched by adequate research efforts include tropical agriculture and water management, aquaculture, and low-cost wastewater treatment.

Difficulty of regional and international cooperation

Improved regional and global management of the environment must complement national policies and investments. For example, the Mekong River and Nile Basin Initiatives have successfully improved the joint management of transboundary rivers and watersheds. Another promising example is the Amazon Cooperation Treaty Organization, which aims to develop a joint management strategy for the Amazon Basin among all riparian countries. Other environmental challenges in need of well coordinated regional responses are desertification, management of coastal and freshwater fisheries, and pollution of water and air. For example, halting eutrophication of Africa's Lake Victoria will only be possible if all shoreline countries—Kenya, Tanzania, and Uganda—cooperate to reduce nutrient loads in the lake.

Improved regional and global management of the environment must complement national policies and investments

Existing regional organizations often lack the mandate or funding to address the complex challenge of improving regional cooperation. A key challenge is overcoming the high costs of coordinating the interests of several countries. A closely related issue is the difficulty of allocating benefits and costs of regional environmental management to individual countries. For example, mechanisms must be found to compensate one country for investments in the improved management of transboundary watersheds that benefit all riparian countries. Similar arrangements should be made for countries that reduce sulfur dioxide emissions and thus acid deposition outside their national boundaries. Such transfer mechanisms are notoriously difficult to set up and require sophisticated institutional mechanisms that are largely unavailable in developing regions.

The same challenges are even more pronounced at the level of international cooperation. With few exceptions, such as the successful implementation of the Montreal Protocol, the international community's track record in implementing international environment agreements is poor. Most conventions, especially those that entered into force during the 1990s, are framework conventions that establish agreed-on goals and general priorities, while leaving details of implementation to each national government's party to a convention. Such agreements

A fully informed citizenry will insist on better performance by the public and private sectors

are not legally binding on the countries that sign, which allows for considerable diversity between the needs and capacities of various countries; the drawback, however, is that means of enforcement are weak to nonexistent.

A key challenge is climate change. This issue can be addressed only through a concerted international response that, building on the entry into force of the Kyoto Protocol, engages the United States and the major developing countries, recognizing the concept of differentiated responsibilities.

A second challenge is the multilateral trade system, which currently includes no provisions to stem trade in natural products, such as timber, that are harvested in unsustainable ways. As a result, certain countries export environmental degradation. China, for example, has recently protected many of its forests, but now imports timber from countries like Myanmar and Indonesia, where most timber is harvested with scant regard for minimum environmental standards. Given the tremendous economic gains that can be made from illegal and unsustainable logging, it becomes extremely difficult for governments to strengthen environmental management without supporting standards set through the international trade system.

A third challenge is to stop the overexploitation of international fisheries resulting from the lack of an international management system. Effectively, the international seas are a "free for all" resource where individual countries compete to harvest key fisheries and lack any incentives to protect the common resource. While no easy solutions exist, this task force discusses in chapter 6 what steps could be taken on a regional level.

Limited public awareness

Environmental sustainability has remained elusive because of limited public awareness. While awareness of the natural environment's importance to human health and well-being and of the consequences of human alteration has grown, it is insufficient to build a broad-based constituency for environmental sustainability. Few understand that their governments not only directly subsidize and thus overcapitalize such sectors as farming, fisheries, and forestry, but indirectly subsidize these same industries by underwriting the remediation of the resulting environmental damage with tax revenues.

A fully informed citizenry will insist on better performance by the public and private sectors that serve them; this insistence will promote more rational, equitable, and environmentally sustainable practices. A major effort must be made to educate the public about the direct connections between the health of their local natural environment and the long-term health and economic well-being of themselves, their families, and fellow citizens. This effort must be country specific and nationally generated. In sum, an informed citizenry may be one of the most effective ways to move expeditiously toward environmental sustainability; while it is an insufficient condition for achieving environmental sustainability, without it, there is little hope that significant change will occur.

2

The solution

Part 2

The solution

As suggested in Part 1, achieving environmental sustainability requires dramatic changes in the ways societies and citizens manage their biodiversity, wastes and byproducts of production and consumption processes, and consumption patterns. This part describes the direct investments and structural changes required at local, national, regional, and global levels. The environmental management measures identified in chapter 4 address the direct drivers of degradation identified in chapter 2. These investments are the basic tools and strategies required to achieve environmental sustainability.

Improving environmental management also requires addressing the indirect drivers—the underlying causes—of environmental problems. These structural changes, described in chapter 5, must complement and occur in parallel with the technical solutions offered.

Few of these proposed solutions are new. Taken alone, they cannot bring about the dramatic shift in modes of production, extraction, transportation, and consumption essential to achieving long-term global environmental sustainability. Taken together—built into the core of efforts to address hunger, health, poverty, and other major challenges—these solutions can put the world on a better path toward environmental sustainability. At all levels, the world community has been most challenged not by designing solutions to problems, but by bringing their solutions to fruition on the ground. To this end, chapter 6 describes how governments, institutions, markets, and other stakeholders can implement the investments and structural changes called for.

The task force urges governments and other actors to recognize the importance of these investments in achieving all of the Millennium Development Goals; it further urges the appropriate actors to take the recommended actions and make the needed investments to stem the rate of environmental degradation.

Environmental challenges are both complex and unique. A multiplicity of institutions must respond to them, and solutions must be adapted to regional and local conditions. This task force recognizes that one size does not fit all; thus, it provides no blueprint for how countries and regions can achieve Goal 7 or quantitative targets for every problem. Instead, it offers suggestions on how to organize the process of integrating the principles of environmental sustainability into all policies and management strategies. It draws attention to implementation challenges and stimulates debate on how to overcome the shortcomings of the existing approaches identified in chapter 3. Ultimately, however, practical solutions will depend on the energy, ingenuity, and commitment of local stakeholders.

Investing in environmental management

The overarching goals of environmental management are to ensure the sustainable delivery of ecosystem services to human societies and manage the wastes and byproducts of production and consumption processes that threaten human and ecosystem health. This chapter recommends specific investments and actions to improve environmental management in each of the six elements of the world's environment identified in chapter 2 as having strong links to the Millennium Development Goals: agricultural production systems, forests, freshwater resources and ecosystems, fisheries and marine ecosystems, air and water pollution, and global climate change.

To integrate environmental sustainability into national development strategies, this task force recommends taking an ecosystem-based approach to guide the planning of interventions in environmental management.[1] This management approach is a strategy for the integrated management of land, water, and living resources that considers the connectivity of different landscape elements. Management activities are based on the protection of critical components of ecosystems and consider the needs and benefits of all stakeholders within the ecosystem. In 2000, the Convention on Biological Diversity adopted it as the primary framework of action for implementation. Two years later, the World Summit on Sustainable Development recognized the approach as an important instrument for advancing sustainable development and poverty alleviation. The ecosystem approach emphasizes the links between human activity and the sustainable functioning of ecosystems (box 4.1). Environmental management must also support conservation of biodiversity in all its forms. Balancing conservation with sustainable use is essential for ensuring long-term environmental sustainability in all countries.

Three major conditions are required to achieve long-term environmental sustainability. First, biodiversity conservation and environmental management

Box 4.1

Ecosystem approach: framework for implementation

The ecosystem approach to landscape and resource management recognizes the complex nature of ecosystems and their relationship to the human populations they benefit. It includes relevant sectors of society in decisionmaking and resource management responsibilities. The approach moves away from discrete, sector analyses of the environment toward a more holistic view of interactions between ecosystem components. It is based on the application of scientific methods appropriate to each level of biological organization and recognizes that humans are an integral component of many ecosystems.

Guidance in implementation

The ecosystem approach seeks to maintain ecosystem functions and services, enhance equitable sharing of benefits, promote adaptive management strategies, implement management actions through decentralization to the most appropriate level, and foster intersectoral and interdisciplinary cooperation.

The IUCN Commission on Ecosystem Management has identified five steps, encompassing twelve principles, to guide environmental managers at all levels in their implementation of the ecosystem approach (Shepherd 2004):

- Determine the main stakeholders, define the ecosystem area, and develop the relationship between them.
- Characterize the ecosystem structure and function and establish mechanisms to manage and monitor them.
- Identify key economic issues that will affect the ecosystem and its inhabitants.
- Assess the ecosystem's likely effects on adjacent ecosystems.
- Determine long-term goals and flexible ways of reaching them.

Sector application

The ecosystem management approach can be applied on a sector basis, as appropriate for the current structure of agencies and organizations that engage in environmental management. Sustainable forest management, integrated water resource management, and ecosystem-based fishery management are examples of effective emerging models that move from narrowly defined sector models toward more inclusive and holistic strategies for environmental sustainability.

must not be confined to protected areas; on the contrary, they must be mainstreamed into all production activities: agriculture, industry, mining, and others. Second, the solutions offered must be integrated across sectors; otherwise, advances in one sector may preclude advances in others (box 4.2). Third, the investments must be implemented in an enabling environment: sound institutions and policies, market equity for the poor and women, and access to information and technologies for all. Without these conditions, management investments will fail to alleviate poverty and reverse environmental degradation.

Element 1: agricultural production systems

Agricultural production systems around the world are increasingly vulnerable to overuse, inappropriate land-use practices, and altered weather patterns resulting from global climate change. Overcoming these drivers of degradation requires increased use of sustainable agriculture techniques, restoration and appropriate management of desertified lands, and protection of surrounding natural habitat.

Box 4.2 Long-term degradation offsets short-term technical solution in Bangladesh

Source: Ali 2004; UNU/PLEC.

When A. M. S. Ali returned to his village on the Bangladesh floodplain to complete a comparative study on the conditions and yield of rainfed and nontraditional rice fields, he found that technical solutions implemented 15 years earlier had caused more problems than they had solved. To avert an impending famine stemming from drought and flood, low-lift pump irrigation had been installed, which had made water more accessible over the short term, allowing for a winter rice crop. In addition, the use of high-yielding variety rice, chemical fertilizers, and pesticides had increased production enough to ensure self-sufficiency of a population that had grown 46 percent over 15 years.

Despite these short-term gains, Ali discovered that the rice fields had undergone long-term degradation. While all land showed evidence of degradation, Ali found alarming differences between traditional rainfed systems (wooden plough, ox manure, no chemicals, and traditional varieties) and nontraditional irrigated systems (groundwater irrigation, power tilling, agrochemicals, and high-yielding variety rice). In the nontraditional fields, soil acidity was higher as a result of dissolved sulfide minerals pumped in groundwater; soil loss and compaction were also higher, and chemical residue was substantial in both soil and water. Under pump-irrigated fields, aquatic life had suffered, and the water table had severely fallen. Following land and water degradation, yields of high-yielding variety rice had fallen dramatically. In sum, the short-term gains in rice yield had been partially offset by long-term degradation of soil and water resources.

[illegible] cropping strategies in response to changing temperatures and precipi- [illegible] adapt to climate variability and change.

Protect and improve soils. Where soil erosion is a concern, production systems should include multifunctional tree crops, such as fruit and timber, perennial crops, and no-till farming techniques. Farmers at all scales of production can reduce farm soil loss and erosion into rivers by maintaining vegetative cover on fields as permanent crops or seasonal cover crops, particularly nitrogen-fixing plants. Systems that include multipurpose trees—those that provide multiple uses, such as fuelwood, timber, and fruits—are especially useful. Where soil fertility is limiting, production systems should include components that provide green and animal manure.

Use water sustainably. Where water scarcity is a problem, farmers should select crops on the basis of water requirements. With appropriate extension advice, farmers should develop alternative water sources and rainwater harvesting and storage techniques.

> **Investments to counter desertification should center on prevention strategies to protect drylands**

Maintain agrodiversity. Farmers should be encouraged and supported by extension workers, government and nongovernmental agencies, and buyers to diversify production systems to include home gardens, crop fields, orchards, livestock, and agroforestry areas, and to maintain high crop genetic diversity to improve food and income security. Government and nongovernmental organizations can support seed banks for local varieties and community seed exchanges to promote the crop diversity.

Mobilize local knowledge and experience. Agricultural agencies (including universities and extension agencies) should work with local farmers under local conditions on research and development of appropriate and sustainable technologies and farming methods for increasing crop productivity. Farmers networking and participatory development in good agricultural practices should be emphasized. Training programs for extension agents should emphasize technical and cultural sensitivity to local needs and conditions, including race and gender issues. These activities must be accompanied by effective systems for mainstreaming and adapting their use to other areas. Agricultural agencies should finance, and in some cases facilitate, farmer-to-farmer demonstration and extension activities. Particular attention must be given to young farmers and recent migrants who may be unaware of the area's productivity limitations.

Improve crop management, storage, and use. Local leaders and civil society organizations can assist farmers to form producer groups to coordinate planting schemes, provide technical and market information, and provide political advocacy at municipal and state levels. In addition, coordinated efforts among producers, extension workers, and buyers should focus on improving on-farm and intermediary crop storage capacity, including methods for drying or processing raw products. To reduce crop wastage and loss, state and local governments should invest in improving feeder roads to improve rural farmers' access to markets.

Restoration and management of desertified lands

Desertification results when mismanagement of soil and water resources leads to salinization of groundwater and soil, erosion of topsoil, and depletion of soil nutrients. Close to 250 million people have been affected by desertification, and 1 billion are at risk. Management investments to counter desertification should center on the adoption of prevention strategies to protect drylands and the mobilization of information and technology.

Adopt prevention strategies to protect drylands. Farmers should take immediate measures to prevent desertification in vulnerable drylands. Dryland protection practices include making optimal use of rainfall through water harvesting

Investments to protect surrounding habitat should focus on rationalizing land-use planning

techniques that maximize retention and conservation measures, safe re-use of treated wastewater for irrigation, tree planting and management and planting of cover crops for arresting soil erosion, improving groundwater recharge, and preventing overgrazing. Farmers and extension workers should sustainably intensify agriculture using technologies that do not increase pressure on dryland water and soil provisioning services. Supported by the Ministry of Agriculture, extension workers should work to understand the knowledge, science, and technologies that underlie successful local systems of managing water and soil in drylands, and then work to implement these and other improved strategies and technologies where needed elsewhere.

Mobilize information and technology. As a preventive and corrective measure, land-use agencies and communities of farmers should plan land-use zoning according to the agroclimatology and the agroecological potential of the land. Information related to soils, climate, water, and other factors should be provided by agencies and organizations that routinely collect, compile, and analyze such data. Governments should strengthen and broaden data collection and analysis efforts through increasing funding; improving human and technological capacity; and facilitating coordination among stakeholders, including the ability to monitor and forecast change. For both planning and information local initiatives could draw on larger initiatives, such as the International Assessment of Agricultural Science and Technology for Development.[2]

Protection of surrounding natural habitat

Sustainable agriculture practices require protection of surrounding habitats and water systems. In addition, more than one-fifth of the household incomes of poor rural communities comes from forest products. Overharvesting, destruction of forest and plant cover, land extension, and overuse of water resources can destroy fragile ecosystems; increase risk of drought, flooding, and erosion; and reduce water supply and quality. Thus, specific investments to protect surrounding habitat should focus on rationalizing land-use planning and setting up communal ownership and management rights systems.

Rationalize land-use planning. Farmers and ranchers should avoid expansion of existing production areas to prevent the destructive extensification of productive activities into natural areas. Education and extension assistance, credit programs, or land-use regulations may be needed as incentives to increase productivity on existing agricultural lands. Government should also assess the links between international market demands and land conversion, as a basis for land-use planning.

Rationalize fertilizer use. To prevent eutrophication and chemical contamination of water bodies and groundwater, farmers at all scales must optimize the

Investments are needed in training, monitoring, and assisting local communities to claim and manage their rights

[illegible] farmers administer fertilizers when they are most needed—so that nutrient input does not exceed crop removal, which [illegible] and coastal water [illegible] ependence and operational effectiveness, direct funding for environmental agencies and strengthening of environment exper- [illegible] the form of grants. To implement the commitments of the Plan of Implementation of the World Sum- [illegible] manage their surrounding resources. Investments in training, monitoring, and assisting local communities to claim and manage [illegible] ity [illegible] government institutions charged with managing the environment must focus on streamlining the number of organizations charged with envi- [illegible] natural resource management. In many countries a maze [illegible] ated ecosystem management of river basi [illegible] ge than in water resource management, which touches virtually every sector of [illegible] lobal Water Partnership emphasizes the importance of adequate coordi- [illegible]

Element 2: Forests

[illegible]

Combating forest degradation requires raising income dramatically in the informal forest sector

[illegible] responsibilities, and raise incomes of rural households. Increased real income of at least 200 percent is both feasible and necessary [illegible] sustainable management.

Improve governance and gender equality

[illegible] engaged in informal forest utilization.

Another critical but frequently overlooked dimension of environmental [illegible] and institutions must engage constructively with local people to ensure that an environment for local action exists [illegible] product management, harvesting, processing, and marketing.

Correcting market failures and distortions [illegible] *of river basin systems—* [illegible]

Governments must work closely with local forest authorities and resource users

services that underpin achievement of all Millennium Development Goals. Such systems represent critical cornerstones for human development and environmental sustainability.[3] To achieve integrated ecosystem management, specific investments should focus on increasing regional coordination, providing technical assistance, and implementing best practices.

Increase regional coordination. Regional authorities, with the participation of governments and other stakeholders, should identify how river basin components benefit stakeholders both directly and indirectly to identify tradeoffs and agree on management actions.[4]

Provide technical assistance. Governments should support landowners and forest managers through incentive systems, credit programs, extension services, and market structure measures that ensure forest systems' long-term delivery of the ecosystem services in demand (for example, timber, nontimber forest products, water, soil stabilization, protection of riparian zones, carbon storage, and recreation). This support involves investment in research and development to develop new technologies, monitoring of progress made to adjust practices through adaptive management, outreach and extension programs, farmer-to-farmer demonstration and training, payment for ecosystem services, and technology transfer.

Implement best practices. Governments must work closely with local forest authorities and directly with resource managers and users to ensure that protection, restoration, harvesting, and regeneration regulations and interventions, including agroforestry practices, are appropriate to the goals of sustainable delivery of ecosystem services and economic development. Proposed activities can be benchmarked against appropriate best management practices, such as the Food and Agriculture Organization's model code of forest practice, certification standards, traditional and indigenous knowledge, fair trade and organic production standards, and standards for soil and water protection.

Protection and restoration of ecologically viable representative areas of all major forest, shrubland, and pasture vegetation types and their biodiversity

These measures are proposed to complement actions related to effective management of the matrix forest area within which protected areas exist. The disappearance of such areas is likely to affect an ecosystem's ability to deliver services to and enhance the well-being of the populations that inhabit and benefit from it. The actions also support the fundamental assumption that no single number can determine the percentage of the Earth's surface that should be placed under protection; this depends on numerous variables, including the nature of the areas and their size, distribution, and viability (ecologically, economically, socially, and politically). To achieve this objective, investments

Restoration activities should be focused on areas of human vulnerability

should center on coordination of conservation strategies, addressing the concerns of vulnerable populations, controlling alien invasive species, and compensating affected stakeholders.

Assess and operationalize conservation strategies. Governments should assess, using existing databases and ground-level techniques, the extent to which major vegetation types are within an effective protected areas network. Countries should then draw up or update plans for each vegetation type to ensure that remaining protection measures are effective. Criteria for effectiveness are that protection measures contribute to conserving the viability and functionality of the ecosystems and ensuring equity of current protection arrangements with respect to current and former traditional and customary landowners. These assessments, including the geographic, ecological, and social components, should be carried out in each country through coordinated efforts of relevant civil society organizations, university students and faculty, and government agencies (with international guidance and data from the IUCN World Commission on Protected Areas or a comparable body).

Restore areas that pose a human safety threat. Forest ecosystems that have been heavily disturbed and thereby put people at risk (such as wetlands, hillsides, and floodplains) should be restored to provide their vital services of retaining rain and flood water and securing soil and land on slopes (see map 2.1). Municipal authorities and local residents in the affected area, with the support of civil society organizations, should lobby for national support of expertise, when needed, and funds. Economic hardships that ensue as a result of restoration, and subsequent protection or management, should be mitigated through a consultative, equitable, inclusive, and transparent process involving all stakeholders.

Control alien invasive species. Restoration activities should support or establish invasive species eradication programs. Control and eradication will require multiple and coordinated approaches, including physical and chemical removal. International trade organizations should establish regulations and appropriate procedures to ensure that potentially invasive species are not transported inadvertently. Programs such as the Water for Work Programme in South Africa can provide local people jobs and improve ecosystem services.

Compensate affected stakeholders. National governments and local communities should invest in nonconsumptive, low-impact uses of the protected areas (for example, ecotourism or carbon storage), with a view to equitable sharing of resulting benefits among those most responsible for ensuring protection or those most affected. Where protection or restoration precludes income-generating activities that previously occurred, implementing agencies must develop mechanisms and arrangements for effective and equitable compensation, if compensation is an

Slowing freshwater degradation requires reducing demand

[illegible] for environmental goods are quotas and permit systems, whereby the permit to consume the resource must

Element 3: freshwater resources and ecosystems [illegible] permits to control

[illegible]

[illegible] contamination from both natural and human-generated pollu-

[illegible] integrated water resources management (IWRM) [illegible] UN Millennium Project Task Force on Water and Sanitation (UN Millennium Project 2005a) [illegible] IWRM strategies [illegible] 2005, as agreed at the 2002 World Summit on Sustainable Development in Johannesburg.

Phase out environmentally damaging subsidies

Reduced demand, especially in cropping systems [illegible] e phasing out of environ-

[illegible]

[illegible] input-intensive practices and excessive degradation. For example, in parts of rural India, heavily subsidized electricity

Increase water use efficiency [illegible]

[illegible] at increasing the cost of water to farmers [illegible]

[illegible] where subsidies are required to reduce poverty, the task force proposes several principles for the design of subsidy schemes.

Identify subsidies that exacerbate [illegible]

[illegible]

Reducing pollution to acceptable maximum levels requires establishing and enforcing targets

[illegible]

technical challenges lies, since experience shows that the design of effective

[illegible]

downstream pollution, and health problems among farmers. The phasing out

[illegible]

contamination (Millennium Ecosystem Assessment 2004a, ch. 7). Chemical

[illegible]

environmental management. While the principle of community management

[illegible]

to the lowest administrative level to ensure maximum responsiveness to par-

[illegible]

Overextraction of upstream surface water and physical alteration of river

[illegible]

Slowing degradation requires managing fisheries at sustainable levels

maintain their biodiversity, investments must be directed toward rationalizing resource distribution and controlling alien invasive species.

Rationalize resource distribution. Scientists must determine the environmental flow—the amount of water that must remain in a river system to maintain ecosystem functions and biodiversity—of rivers and requirements for ecosystem function in coordination with local people, who often have longitudinal knowledge of hydrological dynamics and conditions. Scientists, with the agreement of local users and the assistance of civil society organizations, should then lobby government to put in place regulations and policies to ensure sufficient allocation of water rights to those ecosystems (box 4.3). Because ecosystem science is inexact, it will be necessary to apply the principles of adaptive management in adjusting consumption levels along the water course.

Control alien invasive species. Government programs should be established with adequate public and private support to control aquatic and riparian invasive weeds (for example, water hyacinth or Australian myrtle) through physical clean-up initiatives similar to the South Africa Water for Work Programme. These could be government-initiated programs, with private sector and civil society organization support.

Element 4: fisheries and marine ecosystems

Human activities are rapidly depleting fish stocks and destroying coral reefs and other critical aquatic habitats. Increasing demand for marine products and services, coupled with degradation of inland watersheds and fishery habitat and excessive capture of fish in many inland waters, are resulting in irreversible losses in the productivity of fisheries and aquatic ecosystems. Slowing degradation requires managing fisheries at sustainable levels, rebuilding depleted fish populations to healthy levels, and establishing a network of representative, fully protected reserves.

Sustainably managed fisheries

Managing fisheries at sustainable levels requires taking an ecosystem-based approach because, as Pikitch and others (2004) argue, robust fisheries depend on healthy marine ecosystems.

Box 4.3
Australia caps water

Australia is already reforming its water allocation system by introducing a water cap in the Murray-Darling basin, with subsequent development of a water market. Water allocations are being traded and incentives provided to increase water productivity. Guidelines must be developed for minimum environmental flow in rivers as baselines for planning and management to protect against overappropriation and facilitate rehabilitation.

Restoring depleted fish populations requires eliminating unsustainable fishing practices

Implement ecosystem-based fishery management. Scientists, fishers, and policymakers must collaborate to develop ecosystem-based fishery management plans, taking into account the precautionary approach and appropriate regulations and institutions to implement the plans. Since their current and future livelihoods depend on resource conditions, local fishers should be consulted in decisionmaking processes that determine the scientific agenda, research implementation, and regulations. Ecosystem-based fishery management can be implemented now, even in cases where relatively little information exists, through the judicious use of the precautionary approach (Pikitch and others 2004). Some elements of the approach are already being practiced, for example, under the auspices of the Commission for the Conservation of Antarctic Marine Living Resources.[5] National governments must take a lead role in coordinating the inclusion of all stakeholders in both decisionmaking and implementation phases. To be effective, the ecosystem-based fishery management approach must align needed structural changes with the cultural and economic needs of local fishing communities and their members.

Restored levels of depleted fish populations

Restoring depleted fish populations to healthy levels requires eliminating unsustainable fishing practices, aligning land and water conservation policies, controlling overfishing, and establishing and achieving biomass targets.[6]

Eliminate unsustainable fishing practices. Global fisheries authorities must agree to eliminate bottom trawling on the high seas by 2006 to protect seamounts and other ecologically sensitive habitats and to eliminate bottom trawling globally by 2010. National governments must enact regulations against poison/cyanide fishing, dynamite blasting, and other destructive fishing practices, and enforce regulations in areas where such fishing is already illegal. Fishers must work with fisheries engineers to develop and implement technologies to minimize by-catch, and local authorities must eliminate extraction of coral for construction and as tourist souvenirs. Recently, scientists, conservation groups, and the UN Informal Consultative Process on Oceans and the Law of the Sea (UNICPOLOS) for the UN General Assembly called for imposing a moratorium on bottom trawling. The immediate moratorium would prevent irreversible destruction on the high seas and provide more time to fully assess deep sea biodiversity, fisheries, and ecosystems; determine their vulnerability to deep sea fishing on the high seas; and adopt and implement protection laws.

Align land and water conservation policies. Land-use authorities must work in collaboration with coastal watershed experts to stop or control off-site activities and land-use practices that damage coastal and marine ecosystems. Using national financing and local incentives, reforestation programs in riparian areas can help mitigate flooding and soil erosion; multiple benefits include providing

A global accord must be reached on mandating reduction of fisheries mortality

[illegible] following a full environmental impact assessment.

Provide public access to information

[illegible] ernments to release information about toxic pollutants and involve citizens in the design of biosafety policies in such countries as Chile, [illegible] national governing [illegible]

[illegible] set the biomass levels at which maximum sustainable yield [illegible] and Padua 2005). In coordination with efforts to meet Millennium Develop- [illegible] ion, training in environmental edu- [illegible] oceans, with a long-term goal of 30 percent, is cons [illegible] age of protected areas.

National targets to reduce exposure to toxic chemicals should follow international agreements

Increase coordination and coverage of protected areas. Designating a network of representative, fully protected and soundly managed marine reserves requires international coordination to develop effective national networks of protected areas and establish marine protected areas in areas beyond national jurisdiction. It also requires the cooperation of fishers and other users of marine resources.

Element 5: air and water pollution

Conversion of natural habitat and related patterns of overproduction, overconsumption, and mismanagement have resulted in environmentally unsustainable levels of air and water pollutants. Actions taken to reduce exposure to toxic chemicals and child mortality caused by indoor air pollution and waterborne diseases provide the double benefit of addressing Millennium Development Goal 4 on child mortality and Goal 7 on environmental sustainability.

Reduced exposure to toxic chemicals by vulnerable groups

National targets to reduce exposure to toxic chemicals—particularly pesticides, heavy metals, persistent organic pollutants, and criteria air pollutants—should follow the standards set by relevant international agreements. Investments should focus on adopting integrated pest management strategies, improving frameworks for chemicals management, and implementing standards for environmental management.

Adopt integrated pest management strategies. To reduce soil, air, food, and water contamination caused by toxic agrochemicals and worker exposure, farmers should widely adopt integrated pest management and other sustainable agriculture strategies and practices, especially in large-scale production systems. Successful examples should then be mainstreamed into agricultural extension services. Producers, especially industrial ones, should follow the Food and Agriculture Organization International Code of Conduct on the Distribution and Use of Pesticides. A simple yet effective example of a best practice for avoiding contamination is to rinse empty pesticide containers at least three times, to use any remaining product before the container is recycled or disposed of.

Improve frameworks for chemicals management. National and municipal government authorities should establish and enforce legislation, policies, and programs to manage chemicals safely throughout their life cycles (including implementation of extended producer responsibility or product stewardship). Education and training on safe chemical management and good environmental practices and the organized and systematic involvement of different sectors of society on policy, legislation, and program formulation and implementation should be developed as a means of creating synergies and outreach. Enforcement of regulations requires training and employment of chemicals experts across all sectors. These efforts should draw on existing and emerging multilateral environmental agree-

Reducing indoor and outdoor air pollution will require investment in cleaner technologies

ments, including the Rotterdam, Stockholm, International Labor Organization, and Basel Conventions and the Montreal Protocol. In addition, governments and industries should support such international policy development strategies and actions as the Strategic Approach to International Chemical Management.

Implement standards for environmental management. The private sector must adopt and promote the development of cleaner production technologies, best environmental practices, environmental management systems, and ISO 14000 standards in productive activities—in particular in developing countries, to support small and medium-size industries and workshops. Special attention should be paid to promoting chemical waste prevention, recycling, and the exchange of discarded materials between industries to avoid their elimination as wastes, as well as to the environmentally sound management of wastes throughout a chemical's life cycle.

Substantial reduction in under-five mortality and morbidity by reducing pollution

Reducing levels of indoor and outdoor air pollution to achieve a substantial reduction in under-five mortality caused by pneumonia and acute respiratory infection can contribute to meeting the targets of Goals 4 and 7. It will require investment in cleaner technologies and raising of public awareness.

Invest in cleaner technologies. Government extension agencies, civil society organizations, and aid organizations should promote efficient stoves that take into account indoor air pollution considerations and should encourage switching from solid biomass to cleaner fuels. Convincing households to make the switch requires considerable technical assistance and subsidies to the poorest households for stove adoption and maintenance. Financial support must be complemented by local capacity building and training in stove maintenance to ensure sustainability.

Raise public awareness. Public and private entities should sponsor health awareness programs, especially those that target women; these programs should be presented using audiovisual media in the waiting areas of health clinics, hospitals, and other types of medical facilities and covered by the public media. In addition, emphasis should be given to ratification of and compliance with the World Health Organization Framework Convention on Tobacco Control, set to become international law in February 2005.

Substantial reduction in under-five mortality and morbidity caused by waterborne diseases

Reducing under-five mortality and morbidity requires actions to improve environmental management of biological wastes, as well as the infrastructure targets and actions associated with poor access to water and sanitation, covered

Sound management of organic wastes will be required to address contamination and methane gas emissions

under Millennium Development Goal 7, target 10. Needed environmental management investments are improved protection of water sources and development of new water-collection techniques.

Protect water sources. Municipal planning and land-use authorities should design and implement, with relevant land users, strategies for protecting natural sources of drinking water from untreated runoff in housing compounds and livestock yards by controlling location of defecation areas and maintaining vegetative barriers alongside water courses. The installation of composting toilets and the environmentally sound management of livestock feces to generate compost or biogas and energy should be considered in planning the protection of water sources.

Develop new water-collection techniques. Local and municipal initiatives must be taken, in conjunction with health authorities, to develop and promote rainwater harvesting, storage, and disinfection technologies.

Reduced methane generation related to biodegradation of organic wastes and its contribution to climate change

Organic wastes are the most common wastes in developing countries and are responsible for the generation of methane gas, which is linked to climate change. Sound management of these wastes will be required to address the issue of contamination and methane gas emissions.

Establish integral and environmentally sound management systems. Municipal authorities should regulate and control the management and disposal of organic putrid wastes, combining measures to convert them to compost in both urban and rural areas for agriculture, to use them as animal feed in the case of food wastes, and to produce biogas and energy to reduce methane liberation to the atmosphere.

Element 6: global climate change

Human activities are changing the Earth's climate and its variability. Accelerated tropical deforestation, continued reliance on biomass fuels, and increased combustion of fossil fuels have released significant quantities of greenhouse gases into the atmosphere. The results are more frequent and extreme weather events—droughts, floods, heat waves, and storms—and sea level rise.

Limiting the long-term increase in global mean surface temperature to 2 degrees Celsius and the rate of change to 0.2 degrees Celsius per decade requires a long-term greenhouse gas stabilization goal of 450–500 parts per million of carbon dioxide equivalent (global carbon dioxide emissions would need to peak at 8–10 gigatons of carbon per year between 2005 and 2020 and the emissions of other greenhouse gases would need to be stabilized), with an equitable allocation of national emission rights and intermediate targets.

Research is needed to improve cost-effective energy technologies

[illegible] this overarching target requires investments in cost- [illegible] adequate. Taken together, renewable energy technologies accounted for just 7.7 percent of total government [illegible] 1987 to 2001 [illegible] Likewise, [illegible] ter efficiency of agriculture, low-cost wastewater [illegible] res and [illegible] learning should apply themselves directly to sustainability goals. Universities can adjust curricula, pedagogy, and man- [illegible] capacity for effective environmental management.

Reasons for the inadequate use and financing of science and technol- [illegible] playing field so that environmentally friendly technologies can be [illegible] developing countries alike.

- Competitions and prizes.

Suppor[illegible] *long-term targets.* International organi[illegible]tional civil society organizations must conti[illegible] [illegible] gases, which will send a signal to gov[illegible] [illegible]hold.

Governments must integrate climate change and variability into national planning

Rationalize sustainable production and consumption patterns. To mitigate carbon emissions, land-use authorities must work in concert with private industry and landholders to reduce deforestation and burning; increase afforestation and reforestation; and improve crop, forest, and rangeland management, with appropriate policy frameworks (for example, land tenure). Individuals in wealthy nations must make climate-friendly choices, such as buying clean-energy vehicles and driving less.

Take adaptation measures

Adapting to global climate change requires mainstreaming climate-change responses and investing in adaptation strategies.

Mainstream responses. Governments must integrate issues of climate change and climate variability into national economic and sector planning (including agriculture, water resources, health, settlements, sea walls, dams, and reservoirs). Such integration requires significant coordination among ministries and other government offices and lobbying and education of policymakers by interested civil society organizations and scientific bodies.

Invest in adaptation strategies. At the local level, farmers must invest in alternative cropping strategies in response to changing temperatures and precipitation patterns. This investment involves improved water-storage capacity in drier areas, protecting groundwater recharge areas, shoreline stabilization, and erosion control. Coastal communities require government and international support to manage the effects of sea-level rise. The Global Environment Facility can provide developing countries grant resources to decrease the costs of climate-friendly technologies and climate-change adaptation. Local communities can develop energy-conservation activities and coping strategies to adapt to climate variability and change.

Chapter 5

Structural changes for environmental sustainability

Solving the environmental crisis facing much of the developing world must include not only direct investments in environmental management (chapter 4), but fundamental changes in the ways institutions and economies operate at national, regional, and global scales (McNeely and Milimo 2004). These structural changes must be implemented in order for countries to effectively integrate environmental concerns into all development plans and sector policies. The task force focuses its discussion on four key challenges:

- Strengthening institutions and governance.
- Correcting market failures and distortions.
- Improving access to and use of knowledge.
- Mobilizing science and technology for sustainable development.

Building on these structural changes, the principles of environmental sustainability must be systematically integrated into all sector policies—a process so critical and complex that the task force has devoted a separate discussion to it (chapter 6).

The structural changes proposed in this chapter counteract several indirect drivers of environmental degradation described in chapter 2 and respond to some of the challenges laid out in chapter 3. They underpin and support direct investments in environmental management and are required to effectively integrate environmental concerns into sector policies. It must be stressed that developing countries and their partners must centrally address the tough political choices outlined below before loss of environmental resources can be reversed. To this end, the task force's recommendations address critical national actors typically not associated with environmental policymaking, such as ministries of finance and planning, as well as senior executive decisionmakers in each country. Their active participation is essential if the structural changes proposed here are to succeed.

Countries must improve environmental expertise available within agencies and ministries

This section does not address the structural changes required to stem two of the most important indirect drivers of environmental change—population growth and poverty—since these factors exceed the task force's scope and mandate. Both drivers are addressed in the report *Investing in Development* (UN Millennium Project 2005e), which proposes practical approaches to eradicating poverty and stresses the importance of ensuring universal access to sexual and reproductive health services.

Strengthening institutions and governance

Most developing country institutions tasked with improving environmental management, such as environment ministries or environmental protection agencies, are weak. They lack adequately paid experts, equipment, and operating budgets to design and implement effective environmental strategies. These institutional weaknesses are particularly glaring at the level of local authorities and provincial governments, where environmental expertise is rarely available to effectively address such local challenges as water pollution or falling groundwater tables. Frequently, available human and material resources are spread thin across many—often competing—ministries and institutions.

In addition, sector ministries and agencies charged with infrastructure development, agriculture, land-use planning, or other activities that affect the environment frequently have little or no access to sound environmental expertise. Most do not employ environment experts or consult regularly with environmental agencies.

To achieve environmental sustainability, countries must reform and strengthen environmental agencies and, perhaps more importantly, improve environmental expertise available within agencies and ministries for agriculture, energy, transport, and water supply. Such institutional strengthening requires increased investments, particularly in human resources, and reform of existing institutional arrangements, including management systems.

Train, recruit, and retain environment experts

In most countries, environmental institutions must recruit substantial numbers of environment experts. Likewise, all sector agencies must retain environment experts who can assess the potential effects of sector investment plans on the environment and propose modifications to support environmental objectives.

To improve retention, salaries must be increased and working conditions improved; otherwise, highly qualified and experienced staff will seek opportunities outside public institutions. In many countries, the large numbers of nationals working as consultants for development agencies and international nongovernmental organizations (NGOs) are a stark reminder that public sector jobs are unattractive and underpaid. These environment experts must be offered working conditions commensurate with their skills and pay that can match opportunities outside the public sector. To this end, the wage bill in

Direct funding for environmental agencies must be adequate and predictable

environment ministries, environmental protection agencies, and other government agencies must rise substantially. This recommendation raises the question of how to finance these investments.

Secure sufficient funding for environmental institutions

To ensure a modicum of independence and operational effectiveness, direct funding for environmental agencies and strengthening of environment expertise in sector ministries must be adequate and predictable. To safeguard the independence of environmental institutions and provide continuous funding for environmental policies, some countries have dedicated revenue sources, such as levies on fuels, water consumption, or tourism. These should be earmarked for environmental institutions and policies to ensure predictable financing.

Where domestic resource mobilization is insufficient to meet the Millennium Development Goals, as is the case in most low-income countries, increased aid will be necessary to finance institutional operating expenditures. In most low-income countries, such funding must take the form of grants. To implement the commitments of the Plan of Implementation of the World Summit on Sustainable Development and the Monterrey Consensus, rich countries must make more financing available to developing countries for environmental sustainability (chapter 6). Only through the effective availability of additional funding can poor countries be expected to strengthen their institutions for environmental management.

Reform government institutions and improve interagency coordination

Reform of government institutions charged with managing the environment must focus on streamlining the number of organizations charged with environmental affairs and natural resource management. In many countries a maze of competing responsibilities complicates the design and implementation of environmental programs. While it may not be feasible to create one institution charged with the management of natural resources and other environmental challenges, it is desirable to reduce the number of competing institutions to the barest minimum. Beyond this streamlining, coordinating mechanisms across institutions can be strengthened or set up.

Nowhere does institutional fragmentation present a greater challenge than in water resource management, which touches virtually every sector of society. The Integrated Water Resources Management Toolbox developed by the Global Water Partnership emphasizes the importance of adequate coordination mechanisms across institutions dealing with agriculture, hydropower, sanitation, and watershed management.[1] Such mechanisms must facilitate the sharing of information and the ability to reach decisions jointly. To this end, they must be equipped with partial executive authority, which requires politically painful concessions of authority from each participating institution. This sort of integrated approach to management should be adopted for

Governments and institutions must engage constructively with local people

other elements of environmental change: forests, climate, fisheries, land, and others.

In addition to increasing their level of investment and clearly delineating their responsibilities, environmental institutions must increase their effectiveness by strengthening management systems, improving human resource management, and building staff capacity (UN Millennium Project 2005e).

Improve governance and gender equality

Closely related to institutional gaps and weaknesses is the need to ensure minimum standards of governance, including fighting corruption, improving the rule of law and bureaucratic effectiveness, economic policymaking, and democratization. Poor governance can adversely affect environment outcomes since sound environmental management requires the effective provision of public goods, which in turn depends on the coordination of many institutions and people. Without strong governance mechanisms, competing interests of agencies and information asymmetries result in numerous opportunities to abuse power—typically to the detriment of the poor and the environment. Corruption in particular can accelerate environmental degradation by misaligning private interests with those of the public.

Another critical but frequently overlooked dimension of environmental governance is the inclusion of marginalized groups in decisionmaking and policy formulation. These stakeholders, including women, the rural and urban poor, youth, and ethnic minorities, are too often excluded from decisions that affect their interests. Local and community perspectives are also frequently marginalized or inadequately accounted for in environmental planning, policy, and project implementation. Lessons from the past several decades of development cooperation show that the impacts of sustainable development programs are enhanced when communities are involved and when their knowledge and needs inform delivery. Governments and institutions must engage constructively with local people to ensure that an environment for local action exists and that policies do not undermine the quality of life or the ongoing, often highly innovative, sustainable development efforts of communities. Strategies to ensure gender equality in decisionmaking and implementation must be integrated into environmental planning for many reasons, not the least among which is the declining availability of natural resources (such as fuel or water), which disproportionately affects women. Many countries require improved governance to address this broad array of concerns (UN Millennium Project 2005b, c).

Correcting market failures and distortions

A major challenge for achieving environmental sustainability is ensuring proper functioning of markets for environmental goods and services, removing market failures, and avoiding environmentally harmful market distortions

To account for the costs of environmental degradation, the task force recommends focusing on physical flow accounts

(chapters 2 and 3). Policy and institutional responses to market failures must align private and public interests through a combination of "sticks" and "carrots." In most cases this requires a careful balancing act between safeguarding the interests of the public without unduly constraining the freedom of private actors. The most important policy instruments to address market failures are to account for the cost of environmental degradation in national accounts, introduce payments for ecosystem services and tax reform, phase out environmentally harmful subsidies, strengthen property and land-tenure rights, and improve the regulatory framework.

Account for the cost of environmental degradation in national accounts

Without environmental accounting, economic decisions at national and local levels will be biased toward excessive depletion of natural resources, higher levels of pollution, and degradation of ecosystems. As discussed in chapter 3, serious technical challenges constrain the implementation of comprehensive green accounts. In particular, the monetary valuation of environmental resources and choice of appropriate discount rates remain controversial. As a result, not even high-income countries systematically adjust their national accounts for the cost of environmental degradation.

Therefore, the task force recommends focusing on physical flow accounts without monetizing environmental resources. These accounts can be used to map the flow of resources, such as water, or the yield of coastal fisheries over time and across users. Several countries in southern Africa have successfully used this approach to strengthen environmental management; an impressive example is South Africa's flow accounts for water and forests, which are used to charge farmers and owners of commercial timber plantation for the water flow they divert (Lange and others 2003). This model is now being replicated in other countries.

Introduce payment systems for ecosystem services and tax reform

A central market failure driving environmental degradation is the absence of markets for ecosystems services and mechanisms for monetizing the damage caused by environmental degradation.

The urgent need to introduce additional payment systems that reimburse people and communities for maintaining important ecosystem services, such as watershed protection, pollination, pollution control, or biodiversity conservation, is clear in principle. The challenge lies in implementation. Costa Rica's National Forestry Financing Fund is a pioneering example of payment programs for environmental services. It provides direct financial compensation to land owners in return for a guarantee that the forests are managed sustainably. The program is credited with reversing a decline in forest coverage and contributing to the livelihoods of thousands of families (UNDP 2002).

Other strategies for paying for benefits derived from ecosystem services include mechanisms for providing direct payments to land owners for sustained

The task force supports the Katoomba Marketplace for developing practical payment systems for environmental services

delivery of freshwater, bioregulation, climate regulation, and protection against hazards (Balmford and others 2002; Ferraro and Kiss 2002; Heal 2000). Still other mechanisms support existing national and international markets for certified products, such as timber, organic foods, fish, shade-grown coffee, nontimber forest products, and sustainable production processes.

National payment systems for forests and other ecosystems should be replicated in other countries. To this end, the task force supports the Katoomba Marketplace, which works with pilot countries in developing practical payment systems for environmental services.[2]

One example of an international mechanism that monetizes ecosystem services and enables international transfer mechanisms is the Clean Development Mechanism (CDM) set up under the Kyoto Protocol. While its underlying principles are correct, the CDM remains problematic since it does not account for carbon sequestration by standing forests. Hence it provides much-needed economic incentives for reforestation without strengthening incentives to maintain standing forests.

In addition to introducing payment systems for ecosystem services, countries can monetize the cost of environmental degradation to affect decisions made by individuals and companies (box 5.1). One approach is to shift taxes toward environmental "bads," such as pollution and resource consumption. Some countries have established pollution taxes on emissions (for example, carbon dioxide or sulfur dioxide), which increase revenues and reduce

Box 5.1

Economic principles applied to integrated water resources management in Colombia

Source: Economic Analysis Group, Ministry of Environment, Housing and Territorial Development, Colombia.

Sound water use is central to natural resource policies in many countries around the world. Economic instruments such as prices, taxes, subsidies, and tradable permits can motivate public and private decisionmakers to pursue objectives that are in the best interests of society.

Under a policy of integrated water resources management, the Ministry of Environment, Housing, and Territorial Development in Colombia has promoted the use of water abstraction charges and water pollution charges. These tools were introduced to rationalize decisionmaking in the production and services sectors and to encourage more efficient use of water, both in the amount of water used and in the amount of pollutants entering the water system.

The water abstraction charges and water pollution charges are designed to generate enough funds to send economic signals to users; pay for water resources planning, administration, and control; and invest in improving water availability and reducing pollution. The charges are also expected to improve understanding of the state of water resources, including water demand and usage trends both nationally and regionally, and to contribute to the formulation and implementation of river basin management plans, while minimizing costs and maximizing environmental benefits.

Regional environmental authorities (Corporaciones Autónomas Regionales) are responsible for the application, billing, and collection of water charges, which are based on a minimum national charge adjusted for region-specific characteristics of water resources availability and quality.

Subsidies that exacerbate environmental degradation must be a policy of last resort

emissions. Such revenues can be used to finance investments in environmental sustainability. Other indirect pricing mechanisms for environmental goods are quotas and permit systems, whereby the permit to consume the resource must be purchased, with governments controlling the issuance of permits to control overall resource consumption. For example, governments can create a market by competitively awarding forest concession permits to logging companies or conservation NGOs.

The difficulty of effectively implementing such changes to the tax code must be emphasized. In particular, the environmental monitoring required to support environmental taxes exceeds the capacity of many low-income countries. Moreover, their relatively small tax take as a share of GNP places additional constraints on the effective reorientation of the tax base. In designing such pricing schemes, it is also important to avoid regressive tax burdens that hurt the poor, who disproportionately depend on natural resources for their livelihoods. Lifeline tariffs or equivalent exemptions can be used for this purpose.

Phase out environmentally damaging subsidies

Achieving environmental sustainability requires the phasing out of environmentally damaging subsidies. This phasing out should be done as quickly as is feasible unless the subsidies constitute essential temporary measures for eradicating poverty. Examples of damaging subsidies include production subsidies amounting to $300 billion for farmers in rich countries (OECD 2004). Though at least one order of magnitude lower, agricultural subsidies are also common in many developing countries, resulting in input-intensive practices and excessive degradation. For example, in parts of rural India, heavily subsidized electricity and water for agriculture are a major cause of overabstraction of groundwater by farmers, resulting in falling water tables. Examples of environmentally damaging subsidies in other sectors abound: subsidized fuel and motorized transport, subsidized fishing fleets, or government support for highly polluting industries.

Such subsidies represent a major drain on public budgets and constrain public spending in other sectors. From an environmental and public finance perspective, they should be phased out completely. In certain cases, however, this decision may run counter to important policy objectives, such as the eradication of poverty (for example, by reducing the cost of key agricultural inputs to raise the productivity of smallholder farmers).

A compromise therefore must be reached between social objectives and environmental protection. In cases where subsidies are required to reduce poverty, the task force proposes several principles for the design of subsidy schemes. First, subsidies that exacerbate environmental degradation must be a policy of last resort. Second, subsidies should be de-coupled from the consumption of natural resources (for example, by using direct income transfers, lifeline tariffs for water and energy consumption, or occupational retraining grants). Third, subsidies should be viewed as a temporary intervention connected to an

Formal recognition of customary land management practices is critical

effective strategy that will retire the subsidy after a reasonable period. Fourth, contrary to the prevailing practice in many countries, subsidies should be effectively targeted to the poor. For example, without competitive transport markets, subsidized fuel will increase air pollution and fail to benefit the poor, who cannot afford motorized transport. It is here where one of the greatest technical challenges lies, since experience shows that the design of effective subsidy schemes can be very difficult.

Applied carefully, these simple principles can support poverty reduction while improving environmental management. For example, supporting agricultural inputs and improved farming techniques for smallholder farmers in many parts of Africa can generate increased agricultural productivity, which reduces demand for new farmland. In Indonesia in the mid-1980s, pesticide subsidies totaling nearly $150 million per year resulted in overuse of pesticides, downstream pollution, and health problems among farmers. The phasing out of subsidies reduced pesticide use, increased government savings, and increased rice production (World Bank 1997). In sum, each subsidy scheme must be designed extremely carefully and contain safeguards to ensure that environmental resources are not put under undue stress.

Strengthen property and land-tenure rights

Collapse of traditional forms of land ownership, whereby communities collectively manage common environmental resources, has led to a deterioration in environmental management. While the principle of community management for critical environmental resources is well established, it can be extremely difficult to implement. In many parts of Africa, for example, community forest management has had mixed results. While it is difficult to generalize lessons learned, the task force suggests the following: Clear national standards must be set to govern land tenure and property rights. They should emphasize minimizing the likelihood that two or more competing property rights regimes may apply to the same plot of land. Of particular importance is formally recognizing customary land management practices and indigenous people's rights. Effective property rights must ensure that small-scale producers of crops, livestock, fish and extractive products have incentives to sustainably manage the natural resources on which their livelihood depends.

To the extent possible, application of national standards should be devolved to the lowest administrative level to ensure maximum responsiveness to particular local needs. Improving land management practices is also difficult in the urban context, particularly in slums (UN Millennium Project 2005f).

Improve national and international regulatory frameworks

In cases where it is impossible or not desirable to rely on market-based mechanisms alone, governments must set clear regulatory standards for generation of emissions and waste and extraction of natural resources. For example, national

The effect of international trade on the environment needs to be examined

or local governments—not market mechanisms alone—should decide the maximum level of permissible air pollution based on public health concerns. Since most regulatory regimes in developing countries are weak and poorly enforced, important investments are required to ensure that such countries have the capacity to monitor compliance with environmental standards and strengthen enforcement mechanisms.

A particularly intractable challenge that requires policy responses along many of the above dimensions is the effect of international trade on the environment and ecosystem services. For example, in Indonesia, rising international demand for timber is accelerating deforestation—largely through illegal logging. In the face of rising demand and prices for timber on the international markets, it has become impossible for Indonesian authorities alone to promote sustainable forest management as called for in international agreements and national policy. To stem deforestation under such circumstances, timber buyers must request certification or other proof of sustainable production processes. Finding an internationally agreed on solution—perhaps as part of a multilateral trade deal—should be of highest priority. Examples of such international trade regulations can be found in the provision of the Convention on the International Trade in Endangered Species.

Improving access to and use of knowledge

Achieving environmental sustainability requires greater investment in the generation and use of scientific, technical, and indigenous knowledge. Countries must gather data and indicators to understand existing conditions. Perhaps more importantly, they must develop the institutional capacity to translate data into actionable information to guide environmental decisionmaking by senior policymakers and the general public. While the needed skills, capabilities, and information products differ in each case, the overarching principle is constant: Better information and greater knowledge capacity can significantly improve the quality of decisions and their environmental outcomes.

Establish mechanisms for science and technology advice

Government policymakers at all levels, including employees of line ministries and local authorities, make decisions that affect the environment both directly and indirectly—often without a firm understanding of the environmental implications of their decisions. Until government officials responsible for major initiatives have access to information, understand the management options, and can analyze potential environmental effects and integrate them into the assessment and approval process for policies and projects, they will be unable to make decisions that support environmental sustainability.

Countries can establish advisory structures to provide the highest levels of government expert advice on environmental issues. These advisory structures should be multidisciplinary, capable of applying scientific and technical

Countries should develop broad-based training programs for civil servants and political decisionmakers

expertise to economic and policy analysis. Scientific advisory systems should also integrate information from a wide variety of sources—government, national academies, local experts, and the international community.

To deliver high-quality services, science and technology advisory systems require sufficient and stable funding, a clear statutory mandate to advise the highest levels of government, access to sound and credible scientific information from government and other data sources, an independent operating budget and budget for policy research, and mechanisms for public accountability (UN Millennium Project 2005b).

There are many models for advisory services, and structures must be tailored to specific national contexts. One promising option is to establish offices of science and technology advice in executive offices dealing with sustainable development issues at each level of government (Juma 2004). Other models include establishing national academies of science to mobilize scientists and academics in each country, or building links with institutions in other countries through such organizations as the United Nations Educational, Scientific, and Cultural Organization (UNESCO), International Council for Science, or InterAcademy Council (UN Millennium Project 2005b).

Train decisionmakers for environmental management

In addition to stronger science advisory mechanisms, successful implementation of science, technology, and environmental policies requires a cadre of capable civil servants trained in policy analysis, science policy, and forecasting techniques (UN Millennium Project 2005b). These capabilities are necessary not only for civil servants in environment ministries, but also those in sectoral or line ministries, the finance ministry, and the central government. Civil servants responsible for major initiatives must be able to analyze potential environmental effects and integrate them into the assessment and approval process for policies and projects. To further strengthen national capabilities for environmental management, these experts from developing countries should be fully involved in major international research programs—the International Geosphere Biosphere Programme, World Climate Research Program, International Human Dimensions Programme, and Diversitas—and such international assessments as the International Panel on Climate Change and Millennium Ecosystem Assessment.

The task force recommends that countries develop broad-based training programs in environmental sustainability for civil servants and political decisionmakers. Such courses should become an integral part of in-service training for government employees whose work affects the environment, either directly or indirectly. Implementing such training programs represents a tremendous organizational and financial challenge, particularly in low-income countries. Adequate program financing and staffing should therefore be integrated into poverty reduction and other national development strategies (chapter 6).

Public awareness campaigns are needed to educate citizens about the environment

In addition, considerable investment should be made within institutions of higher learning to develop the next generation of environmental scientists, particularly in areas of energy use, climate change, ecosystem management, population and evolutionary biology, and molecular genetics. A sufficient number of research scientists in these areas will be essential to maintain the quality of scientific information prepared for a new, highly trained generation of civil-service environmental managers and policymakers.

Provide public access to information

Public access to information on local and global environmental issues can promote better policies for environmental sustainability. In 1992 this view was affirmed in Principle 10 of the Rio Declaration on Environment and Development and a decade later reaffirmed in the Johannesburg Declaration on Sustainable Development and Plan of Implementation. Principle 10 articulates public access to information, participation in decisionmaking, and access to justice as key principles of environmental governance. By increasing transparency and accountability, public involvement in decisionmaking can improve the quality of environmental governance and enhance long-term sustainability of solutions. In Johannesburg, for example, the official Partnership for Principle 10 was launched with broad support and has scored important successes in encouraging governments to release information about toxic pollutants and involve citizens in the design of biosafety policies in such countries as Chile, Thailand, and Uganda (Partnership for Principle 10 2004).

Public awareness campaigns are needed to educate citizens about their right and responsibility to promote environmental sustainability in both public and private sector activities. In developing countries, civil society organizations should be supported financially to develop and implement awareness programs using mass media and public workshops. These tools, which highlight environmental conditions, drivers, and responses, can help citizens understand the links between human behavior and the environment. These campaigns should also emphasize, particularly to wealthy citizens, that habits of excessive consumerism must be curbed.

Moreover, environmental issues should be integrated into educational curricula at all levels—from primary schools to universities—and made concrete through practical internships and hands-on field experience (Pearl, Buchori, and Padua 2005). In coordination with efforts to meet Millennium Development Goal 2 on universal primary education, training in environmental education should be provided for primary school teachers and those in charge of informal education programs.

The basic principle underscored is this: people well informed of the importance of a healthy natural environment to the quality of their daily lives will act as important stewards and managers of their environment and the goods and services it delivers (box 5.2).

Box 5.2

Achieving tangible results in pollution control in the short and medium term in Mexico through public involvement

Source: Cristina Cortinas, personal communication, October 2004.

> We need armies of good-willed, informed, organized, and committed citizens, giving support to government authorities to put into place the plans to comply with the Millennium Development Goals.
>
> —Cristina Cortinas, scientist, educator, policymaker, community activist

Public awareness and involvement helped to advance environmental sustainability in pollution control in Mexico through multiple strategies:

Stronger legislation. Citizens of the State of Queretaro helped to develop new laws and regulations on waste prevention and integrated management and, through collective action, helped to fill regulatory gaps by developing, disseminating, and investigating the application of regulations. Establishment of a statewide intersectoral waste management network (www.reqmar.org) helped the government develop state and local waste regulations and promote research projects on environmental standards for managing nonhazardous waste.

Public participation. The new legislation establishes the shared responsibility of all sectors of society for waste prevention and integrated management. Knowledgeable citizens can receive support from the federal Environment and Natural Resources Ministry to conduct training courses for various sectors of society on application of the new legislation. These activities will be linked through the Internet and will establish their own websites to share experiences and exchange information.

Waste minimization. The new legislation focuses on waste recycling and the use of wastes as alternative fuels. It offers incentives for materials sharing among firms to minimize waste. The idea is to link waste management projects with poverty alleviation projects (involving waste scavengers in recycling formal productive chains, for example) and sustainable agriculture (using organic wastes as compost and establishing management plans for empty pesticide containers) and to involve the academic sector in projects to help industries reduce waste generation and increase recycling.

Education. As a waste generator, the education sector has to comply with the new legislation and develop waste management plans that include waste prevention and recycling. This obligation presents an ideal opportunity for educating students by involving them in the design and implementation of the plans and making them aware of the need for waste prevention and recycling in other areas of their lives.

Demonstration projects. Following a community approach and working through the intersectoral network on environmental waste management, projects will focus on activities at the community level. Federal, state, and municipal authorities and other key actors (including waste generators, producers, managers, academic institutions, and civil society organizations) will join in designing and implementing projects. This will reduce the cost burden for any individual stakeholder of developing projects.

Knowledge networks. Because government lacks the capacity to guide citizens in all aspects of preventing and managing waste in an environmentally sound manner, it is critical to establish mechanisms for providing citizens with access to information on how to solve concrete problems.

Capacity development initiatives should be expanded to focus on environmental management

Strengthen global scientific assessments

Such global scientific assessments as the International Panel on Climate Change and Millennium Ecosystem Assessment have proven successful at developing a scientific consensus on key environmental issues. As such they are indispensable for developing sound environmental policies based on the best available science. These assessments can be strengthened in two ways. First, they should be expanded to review regional and subregional environmental challenges. The Millennium Ecosystem Assessment has pioneered this approach through its subglobal assessments. Similar assessments should be implemented for other regional challenges. Second, these assessments should be carried out in other areas, such as agriculture.

Improve extension training and services

Agricultural extension is the bridge between the laboratory and the field, where extension workers can provide information on new cropping techniques and use of inputs, as well as advice on record-keeping, marketing, and managerial skills. Extension workers are also conduits for information flows from farmers to the government, expressing grievances and concerns to public agencies (Birkhaeuser, Evenson, and Feder 1991).

Yet trained extension staff are in short supply, and training methods are outdated. Current methods do not adequately integrate environmental considerations into the content of agricultural extension practices. Preservice training institutions must be improved and expanded, and existing practitioners should receive appropriate in-service training, with sensitivity to local culture and ecology. Extension agents should introduce new ideas and technologies, but should also work to understand successful local production systems and be able to facilitate the scaling up of these systems. Understanding and integrating the knowledge and technologies of local farmers and communities into extension services is a critical ingredient for success.

Governments and donor agencies should support civil society organizations and NGOs to systematically document traditional and local environment-related knowledge, especially in the fields of agriculture and natural resource management, and create channels for farmers or fishers to exchange knowledge and technologies. Extension agencies should fully engage in these efforts. Capacity development initiatives should be expanded to focus on environmental management by identifying smallholder practices and knowledge as models for integrating environmental sustainability with economic development.

Mobilizing science and technology for sustainable development

A frank assessment of the indirect drivers underlying environmental change paints a grim picture. Rising per capita income and consumption coupled with population growth lead to rapidly rising demands for ecosystem services and cause growing levels of waste and pollution. For many countries, it is impossible

Science and technology must be at the center of any strategy for environmental sustainability

to maintain rapid per capita growth in water consumption amid high population growth and increasing demand for food when internally renewable water resources are limited. Likewise, the carbon intensity of all economic activities must fall sharply to stabilize greenhouse gas emissions. In short, current patterns of consumption and production are unsustainable in the face of the finite capacity of ecosystems to produce goods and services and to withstand pollution.

In certain parts of the world, the situation is so serious that one can be forgiven for concluding that environmental sustainability is unachievable. However, this indulgence must be quickly followed by efforts to find solutions, even for the most degraded environments.

Investments in science and technology can dramatically raise the efficiency of natural resource use for all economic activities. They can also complement noncoercive efforts to bring down fertility rates and population growth by moving toward sustainable consumption and production patterns (UN Millennium Project 2005e). Moreover, science and technology can improve environmental monitoring and early warning systems for natural disasters using remote sensing and other IT-based systems. They can also contribute to strengthening ecosystem management and services. For example, simulation tools can improve management of water reservoirs, while remote sensing can strengthen forest management and reduce the cost of enforcing environmental standards and regulations.

Skeptics argue that even the best science cannot generate efficiency gains that can lower natural resource consumption to environmentally sustainable levels. However, this skepticism is speculative and largely unfounded. For example, no one would have anticipated in 1961—the year the first microchip was introduced—that computational power would double every 18 months. In 1971 Intel's leading microchip held just 2,250 transistors; by 2003 that number had increased to 410,000,000.[3] Likewise, whereas the typical car in the early 1970s averaged only 13.5 miles per gallon (17.5 liters per 100 kilometers),[4] today's most fuel-efficient models can travel more than 60 miles per gallon (3.9 liters per 100 kilometers).[5] Such agricultural techniques as agroforestry can triple food output of smallholder farmers without the environmental degradation typically associated with agricultural intensification (Sanchez 2002). Advances such as these were made possible through decades of high investments in research and development. There is no reasonable explanation why similar or greater scientific advances cannot be made over the coming decade to move the world toward environmental sustainability.

Science and technology must therefore be at the center of any strategy for environmental sustainability. Of course, the hope of future advances in science and technology is no substitute for careful stewardship of the environment today. Similarly, such advances are no panacea and present risks, as amply demonstrated by the controversy surrounding genetically modified organisms. These dangers are real and must be carefully managed. Careful impact

Universities and other institutions of higher learning should apply themselves directly to sustainability goals

assessments and broad public dialogue are therefore a prerequisite for increased investments in science.

At the same time, current efforts to promote science and technology for environmental sustainability are woefully inadequate. Taken together, renewable energy technologies accounted for just 7.7 percent of total government energy research and development funding from 1987 to 2002.[6] Likewise, research into increasing the water efficiency of agriculture, low-cost wastewater treatment for cities in poor countries, land management techniques to combat and reverse desertification, or control of invasive alien species is insufficient.

As a first step toward strengthening the generation of knowledge, universities and other institutions of higher learning should apply themselves directly to sustainability goals. Universities can adjust curricula, pedagogy, and management to deal with local issues related to environmental sustainability. In urban areas, for example, more research could be directed toward solutions to such challenges as sanitation and the poor living conditions of slum dwellers. One promising approach involves the study of the urban ecosystem, its ecological cycles, and connection to other ecosystems (Berkowitz, Nilon, and Hollweg 2003). In rural areas, universities could emphasize research, training, and outreach on natural resource management (Juma 2004). As pools of scientific and technological expertise, universities are well positioned to contribute to the expansion of knowledge capacity for effective environmental management.

Reasons for the inadequate use and financing of science and technology, explored in chapter 3, include market failures that constrain for-profit research by the private sector, insufficient awareness among policymakers of the opportunities created by modern science, and insufficient funding. Science and technology for sustainable development can be promoted in at least four practical ways:

- Direct public financing of research.
- Competitions and prizes.
- Precommitment purchase agreements.
- Transparent regulatory standards.

Under direct public financing of research, rich countries mobilize scientific capabilities to address the problems facing the poor (UN Millennium Project 2005e). The UN Millennium Project (2005b) provides practical recommendations on how countries can strengthen their research capacities. To finance major investments in national research systems, poor countries may require increased official development assistance.

A second approach, ex post prizes, is useful for solving well defined problems, such as developing a new technique to combat an invasive alien species (Masters 2002). The international community can administer these prizes. For example, a prize could be set up to reward developers of the first solar cell that can generate a kilowatt-hour of electricity at less than a few cents or reduce the cost of secondary wastewater treatment to a certain threshold.

Transparent regulatory standards can promote research and technology development

A third approach proposed by Kremer (2002) is pre-commitment purchase agreements. These are binding public commitments to buy a product, such as an improved cook stove, at a minimum price. They enable private companies to plan for a minimum production level, thus removing an important element of risk in the development of product-stage research or bulk production of technologies such as improved cookstoves or low water use toilets.

Closely related is the strategic use of government procurement to promote research and development for environmental sustainability. By setting clear minimum environmental standards and safeguards that a supplier must ensure, governments can create a larger market for technologies supporting environmental sustainability.

Finally, transparent regulatory standards for air and water quality, generation of waste, or other parameters can be used to promote research and technology development if set clearly and announced in advance to give companies and individuals time to react to changes in their incentive structure. An example of effective regulation is the European Commission–proposed Registration, Evaluation, and Authorization of Chemicals (REACH), which would require manufacturers to prove that chemicals are not harmful to humans and the environment.

These four options for immediate action can help unleash the potential of research and technology to increase efficiency of resource consumption and production and minimize pollution and other waste. Thus all countries and the international community should implement these options to improve opportunities for achieving environmental sustainability.

Chapter 6

Implementing strategies for environmental sustainability

Achieving environmental sustainability depends on successfully implementing the recommendations described in chapters 4 and 5. This poses a tremendous challenge, particularly in developing countries. Like gender equality, environmental sustainability cuts across line ministries, fields of knowledge, and geographical scales. Only an integrated approach to aligning sector policies and investment strategies with environmental objectives can yield the desired outcomes. Such an approach requires strong political will and a detailed understanding of the complex links between economic activity and the environment. It also necessitates intensive cooperation and coordination among diverse stakeholder groups, mobilization of a variety of technical expertise, and a long-term perspective. These challenges have rarely been met in developed or developing countries.

Many recommendations outlined in this report are not new. Indeed, some were called for at the 1972 Stockholm Conference on the Human Environment and form part of the major multilateral environmental agreements of the past three decades. Despite this long history, most recommendations have not been systematically implemented. It is therefore in their implementation where the greatest need for innovation lies. Practical implementation mechanisms must act at the right spatial scale; bring together the appropriate expertise and stakeholders in a collaborative process; and differentiate between short-, mid-, and long-term needs. This chapter proposes such mechanisms to enable countries to implement the recommendations outlined in chapters 4 and 5.

Of central importance are implementation mechanisms that address environmental challenges at scale through government interventions. Few success stories for improved environmental management provide answers at the scale required to meet the Millennium Development Goals. Mechanisms used to implement small-scale projects will not be immediately applicable to

Implementation mechanisms must bring together expertise from a broad range of fields

countrywide programs. Only if all levels of government—from local authorities to national governments—integrate the principles of environmentally sustainable development into their policies and investment strategies can change occur at the right scale. The task force therefore focuses its discussion here on implementation mechanisms involving governments. This does not mean, however, that civil society and the private sector do not have critical roles to play; the contributions of both groups are critical for environmental sustainability to be achieved.

Since environmental challenges act at local, national, regional, and global scales, corresponding implementation mechanisms must be developed. This chapter distinguishes three levels:

- National mechanisms that develop and support interventions at the national and local level.
- Regional processes to manage transboundary ecosystems or pollution.
- Global mechanisms that address such global challenges as climate change, effects of international trade, and overharvesting of fisheries.

The task force emphasizes the need for national-level implementation, not because regional or global challenges are less pressing, but because it is at the national level where rapid progress must be made to meet the Goals. Unless environmental policies in countries move to the center of national development strategies, substantial progress at regional and global scales will be difficult. Thus, national-level implementation is key to addressing local challenges and laying the foundation for successful regional and global implementation.

In addition to acting at the right geographical scale, implementation mechanisms must bring together people with a broad range of skills and expertise, including representatives from most line ministries, civil society organizations, representatives of local communities, and researchers from diverse fields. The breadth of expertise required for environmental policymaking requires well-functioning institutions and consultative processes, which pose a major challenge for many low-income countries where institutions tend to be weaker. Thus, as discussed in chapter 3, the task force stresses the importance of adopting a pragmatic and gradual approach to investing in environmental sustainability in parallel with strengthening institutions, expanding human resources for improved environmental management, and reforming policies. The focus should not be on getting things 100 percent right, but on getting started.

A major implementation challenge lies in identifying, evaluating, and addressing tradeoffs and synergies between sectoral strategies and environmental objectives. While win-win strategies have received much attention in the policy literature, the more common and intractable challenge lies in finding the right balance between the needs of the environment and short- or medium-term social and economic imperatives. As shown below, identifying and managing tradeoffs requires political processes that senior decisionmakers must lead with the full participation of all stakeholders.

Implementation mechanisms must develop strategies that can address long-term change

Finally, implementation mechanisms must develop strategies that can address long-term change. As discussed in chapter 2, such challenges as climate change or desertification are the result of direct and indirect drivers that act slowly and can only be addressed over the long term. For example, changing the energy intensity of a national economy or increasing the water efficiency of agriculture requires strategies and investments that will show results only if pursued over many years. In addition, implementation mechanisms to achieve environmental sustainability must address possible tradeoffs between short-term and long-term strategies. Many strategies that can yield substantial short-term benefits, such as intensified agriculture in regions experiencing water scarcity, may prove unsustainable over the long term because they exceed the carrying capacity of ecosystems. Similarly, investments to reduce greenhouse gas emissions impose a cost on today's generation, with the benefits from a more stable climate accruing at some future time. No tested institutional models yet exist to deal effectively with intergenerational tradeoffs, even though they lie at the heart of the implementation challenge.

National implementation mechanisms

Most countries have one national development framework, such as the poverty reduction strategies and Poverty Reduction Strategy Papers (PRSPs) common in many low-income countries. Countries should systematically align their development frameworks with the Goals as operational targets. The UN Millennium Project refers to such development frameworks as MDG-based poverty reduction strategies (UN Millennium Project 2005e).

Shortcomings of national strategies

PRSPs and equivalent development frameworks in other countries tend not to address environmental sustainability systematically. For example, Bojö and others (2004) show that few existing PRSPs contain practical strategies for achieving Goal 7. Those that do, focus almost exclusively on providing access to water supply and sanitation. Virtually none pay adequate attention to sustainable forest management, prevention of land degradation and desertification, pollution reduction, or the other key elements of the world environment described in chapter 2. These weaknesses are part of broader shortcomings in the content of PRSPs or similar development frameworks and the process by which they are developed (UN Millennium Project 2005e). Four salient challenges are:

- *Level of ambition.* Even where PRSPs mention the Goals, targets and milestones often fall short of what is required to meet them. All too often PRSPs ask what can be achieved within the confines of current financial, institutional, and other constraints instead of focusing on what it takes to achieve the Goals.
- *Time scale.* PRSPs typically cover a three-year period and are rarely developed as part of a long-term development strategy. This time frame

Environment strategies must be better integrated with line ministries' sector strategies

is at odds with the need for longer term strategies that are required to achieve environmental sustainability.

- *Scope and integration of crosscutting issues.* Few PRSPs cover the full range of Goals. Even fewer adopt an integrated approach to addressing cross-sector issues such as the environment, gender equality, or urbanization.
- *Lack of expenditure frameworks.* Because PRSPs tend not to include detailed budgets or expenditure frameworks, they are difficult to implement. Where budgets exist they are sometimes less than the minimum necessary to implement even the strategies laid out in the PRSPs.

As has been widely documented (Oxfam 2004), the process of developing poverty reduction strategies tends to be inconsistent with developing broad-based, goal-oriented strategies for achieving the Goals, which are nationally "owned." Typically, most work is carried out by planning ministries that have limited environmental expertise and consult poorly, if at all, with other line ministries. Expenditure ceilings are sometimes set for each area before the needs for meeting the Goals are assessed. Hence budget allocations are often woefully insufficient. All too often, the environment ministry is not represented at the key discussions and has little influence on strategy design. Moreover, engagement with civil society and other stakeholders is often insufficient.

The task force recognizes that the shortcomings of poverty reduction strategies partly result from the extreme financial and policy constraints the international community imposes on governments of low-income countries. As a result, few low-income countries today have the opportunity to design comprehensive strategies for achieving all Goals. Expanding the financial envelope available to well governed countries to support MDG-based poverty reduction strategies is therefore a central recommendation of the UN Millennium Project. Financing constraints are compounded by poor communication among ministries and a lack of political commitment to achieving environmental sustainability.

Gaps in current PRSPs are not filled by dedicated national environment strategies. Many developing countries have prepared National Strategies for Sustainable Development or national action plans as part of their obligations under the international environment conventions or other commitments, such as the United Nations Conference on Environment and Development, Convention on Biological Diversity, or Convention to Combat Desertification. While many are excellent, they are rarely implemented. To become effective implementation mechanisms, these strategies must be better integrated with line ministries' sector strategies and receive sufficient financing. To this end, they must become an integral part of MDG-based poverty reduction strategies.

Integrating the environment into poverty reduction strategies

The UN Millennium Project recommends that countries develop one integrated development framework to meet all Millennium Development Goals. This MDG-based poverty reduction strategy can be developed on the basis

The MDG-based poverty reduction strategy must be developed through an open consultative process

of existing PRSPs, National Strategies for Sustainable Development, or other frameworks. The name of the mechanism does not matter as long as it provides an integrated operational framework for implementing and financing strategies to achieve all Goals and ensure appropriate monitoring of progress. The main innovations required to integrate environmental sustainability firmly into poverty reduction strategies involve strategy development, financing, and monitoring. MDG-based poverty reduction strategies could cover three to five years, but must be integrated into a 10-year framework for action that details longer term strategies through 2015 (UN Millennium Project 2005e). This approach is particularly relevant since long-term environmental goals often require long-term actions.

The MDG-based poverty reduction strategy must be developed through an open consultative process involving all key stakeholders. The task force recommends that each country convene an MDG planning group chaired by the national government, but also including bilateral and multilateral donors, UN specialized agencies, provincial and local authorities, and civil society leaders, including often-underrepresented women's organizations. The planning group can then organize a series of thematic working groups, each with broad participation, to develop scaled-up strategies in such areas as health, rural development, and environmental sustainability. Each thematic group should include environment and gender equality experts. Health, education, and environment line ministries must take lead roles in the respective working groups. Development partners and civil society leaders should designate focal points for each of these government-led groups, joining the problem-solving process at the outset, rather than midstream or after the fact.

The working group on environmental management would be tasked with:

- Operationalizing target 9 by identifying specific time-bound targets together with appropriate indicators to facilitate monitoring.
- Identifying policy and investment needs to improve the management of environmental resources, as discussed in chapter 4.
- Appraising the environmental effects of strategies proposed by other thematic working groups and their consistency with environment objectives.

Operationalizing the target at the national level

To guide not only environmental strategies, but also investment priorities across all sectors, operational targets for the environment must be adopted through a political process involving the highest level of decisionmaking (box 6.1). This process should be initiated by the working group on environmental sustainability. First, it should ensure that all relevant line ministries are well represented. Second, key stakeholders should be consulted. These might include NGOs, religious groups, and the commercial sector. The process of consultation should be open and transparent to ensure that the resulting

Box 6.1

Why are operational environmental targets important?

Lack of environmental targets at national and international levels impedes progress toward achieving environmental sustainability. First, whenever environmental needs compete with other development objectives, the latter tend to define policy actions—often at the expense of the environment—since the environmental cost is not made transparent.

Second, without specific outcome targets, it is impossible to track progress or evaluate the success of national programs; this in turn hinders learning and the gradual improvement of environmental strategies.

Third, without operational targets, it is virtually impossible to justify substantial investments in environmental management. This is a critical reason why the budgets of environment and associated ministries are so small.

Fourth, lack of clear national targets contributes to fragmented responses to the environmental crisis. In most countries, multiple organizations at various levels (from the local community to the national government) work on environmental issues. Rarely are these efforts bound by a common vision and goal; as a result, workers' talent and commitment are not fully harnessed.

An MDG-oriented approach to environmental policy requires clear priorities and objectives at national and international levels. Once operational targets have been set, many sources of knowledge are available to guide how they are to be achieved. These knowledge sources include experience, science, and traditional knowledge. The Millennium Ecosystem Assessment confirms this finding and contains detailed sections on available responses to meet environmental challenges (Millennium Ecosystem Assessment 2004c).

targets have the support of all key decisionmakers. Drawing on the best available science, the target selection team should begin with a relatively long list of potential targets, treated as options with cost-benefit information for each.

To be effective, such operational targets must meet various conditions. First, they should be easy to monitor so that progress can be tracked and mid-course adjustments made as necessary. Second, targets should distinguish between short-, medium-, and long-term objectives to be achieved by 2007, 2015, and 2050, respectively. Third, country-level targets must distinguish clearly between outcome targets (such as number of hectares of degraded land to be restored, maximum permissible nutrient load in critical freshwater ecosystems, or protection of key coastal habitats) and process targets (such as adoption of a national strategy for biodiversity conservation). Finally, national targets should draw as much as possible on existing international agreements, such as the Johannesburg Plan of Implementation or similar multilateral agreements.

While science can help inform choices about such targets, the amount of information is overwhelming and requires a sound framework within which it can be fit. For example, the work of the Millennium Ecosystem Assessment (2005c) shows that, based on the best science alone, it is difficult to establish specific targets that must be met to maintain ecosystem sustainability. In part, this conclusion is a call for dramatically improving knowledge of the natural world; however, it also emphasizes a fundamental challenge: Even a

Reaching agreement on environment objectives will require a political process

combination of the best data and the best science is unlikely to provide definitive answers to what environmental sustainability should mean.

As the final decisionmaking and coordinating group overseeing the preparation of the MDG-based poverty reduction strategy, the MDG planning group must review and approve environmental objectives identified by the working group on environmental management. It is within this context that the tradeoffs between environment objectives and exigencies in other sectors must be discussed openly to reach a common position across all working groups. To this end, the relevant experts must clearly lay out the implications of adopting a specific target.

Reaching agreement on environment objectives will likely require a complex political process involving substantial negotiation; only in this way can the necessary compromises among competing needs and interests be brokered. The involvement of civil society organizations is critical for the process to proceed (box 6.2). Once approved by the planning group, the environment objectives must guide the work of all thematic working groups. Each sector strategy should be carefully reviewed to assess whether it is compatible with the environment objectives.

Estimating needs

In a second step, the working group on environmental sustainability should oversee the preparation of a detailed needs assessment to quantify human and financial resources, as well as infrastructure and equipment needed to meet specific environmental objectives agreed on by the MDG planning group (box 6.3). Such needs assessments should provide a rigorous answer to the question "What will it take to achieve the Millennium Development Goal?" In many countries, this needs-based approach will mark an important departure from current practice, which focuses primarily on the marginal expansion of services and investments, with little regard for medium- and long-term objectives. Based on this needs assessment, investment strategies can be developed to improve environmental management (chapter 4), strengthen institutions, build human and technical capacity, and improve policies and regulatory frameworks (chapter 5).

In addition to estimating financial resources, needs assessments should also quantify the human resource requirements critical for achieving environmental sustainability. Based on a long-term assessment of the types and numbers of professionals required to implement strategies for environmental sustainability, countries can develop approaches to training sufficient numbers of people. In this way, capacity constraints can be identified and addressed early in the process of scaling up interventions.

Quantifying the resources needed to meet Goal 7 is a daunting task since environmental conditions, investment priorities, and available interventions are highly country specific. In comparison, investments needed for primary education, health, or water and sanitation can be estimated more easily. The

Box 6.2
What civil society can do

Strategies for achieving environmental sustainability will likely fail without the support and involvement of civil society organizations, including community-based organizations, NGOs, think tanks, social movements, religious organizations, grassroots and indigenous people's movements, and voluntary organizations. These organizations play a crucial role at both international and national levels in achieving the goal of environmental sustainability as they both reflect and respond to the needs of a broad range of constituents and communities.

International level

At the international level, civil society organizations can mobilize and build public awareness of environmental challenges, deliver direct services, and share knowledge of best practices. NGOs have already played a crucial role in raising public awareness of environmental challenges, and pioneering debates on solutions. These groups have often catalyzed international public opinion and mobilized political pressure around important global issues, such as climate change, international trade in endangered species, and deforestation. They have helped ensure that governments recognize and invest in developing sustainable solutions to these challenges.

National level

At the national level, civil society organizations can contribute in at least four ways:

Public advocacy. Civil society groups are essential to create and mobilize grassroots demands that hold leaders accountable and place the Millennium Development Goals at the heart of national development priorities. Strategic alliances with local authorities, national governments, and the international community can raise public awareness of their government's commitment to the Goals, highlight urgent development priorities, and ensure that the needs of diverse groups are taken into account. For example, in 1995, the Centre for Science and the Environment in New Delhi launched an advocacy campaign for tighter controls on air pollution in major cities. Participants published a research report on vehicular pollution, followed by a series of high-profile public meetings with scientific experts, politicians, doctors, university students, and economists. Artists, literary personalities, sports celebrities, and journalists joined the coalition, attracting even more attention. Soon, India's supreme court directed the government to draw up an action plan for controlling air pollution; the resulting strategy introduced vehicular emissions tests, better mass-transit infrastructure, and buses fuelled by compressed natural gas (Pandita 2004).

Design of locally adapted strategies. While the government must lead policy responses to environmental challenges, civil society is crucial in translating these policies into practical solutions. Community groups can help to ensure, for example, that conservation efforts are supportive of their livelihoods, and, where there is a threat to rural livelihoods, that adequate compensation is provided and that investments to restore degraded ecosystems are implemented in cooperation with local populations. An interesting example is the Rehabilitation of Arid Environments Charitable Trust in Kenya, which has worked for more than 20 years to rehabilitate grasslands in the Lake Baringo watershed by setting up private and communal fields protected from grazing animals. Within three years, community management transformed severely degraded terrain into productive land.

Implementation of scaled-up strategies and investments programs. Implementation of large-scale environmental investment programs requires civil society action. In some

(continued on next page)

Box 6.2
What civil society can do
(continued)

cases, the government can lead implementation; in others, civil society organizations must design and implement programs. Most large-scale interventions require community involvement, including public education, dialogue, and national-scale action. Scaling up involves more than simply encouraging the expansion of successful local initiatives. It also requires understanding the factors that have led to success at the local level and taking community perspectives into account in both policy formulation and project implementation. For this reason, governments, NGOs, and other players must work with local people to understand how they have managed to overcome fundamental challenges and carve out sustainable livelihoods. Scaling up also requires that those responsible for setting and influencing policy have a constructive working relationship with community actors and implement policies that promote local leadership and innovation. In 2004, the Nobel Peace Prize was awarded to Wangari Muta Maathai, a Kenyan woman whose leadership of the Pan-African Greenbelt Movement demonstrates the power of community mobilization. Her efforts have helped poor women organize to fight desertification and environmental degradation by planting more than 30 million trees.

Environmental assessments, monitoring, and evaluation. At the national level, NGOs can monitor and implement national commitment to meeting Goal 7. This includes monitoring government policies, private sector decisions (especially in natural resource–based industries), and global policies on the country's environmental outcomes. The Access Initiative (TAI), a global partnership between the World Resources Institute and more than 50 civil society organizations in 25 countries, works to increase public access to information and participation in environmental decisionmaking key themes in Principle 10 of the 1992 Rio Declaration on Environment and Development. TAI has developed common standards and tools for assessing performance and identifying gaps in public participation in environmental governance. This common international method is used to conduct independent national-level assessments of law and practice, holding governments accountable to commitments made in Principle 10. In Chile and Thailand, TAI assessments convinced governments to collaborate with NGOs in developing a Pollutant Release and Transfer Registry.

UN Millennium Project has conducted needs assessments in various low-income countries to quantify these investment needs. For lack of national priority objectives as well as data, these assessments do not include resource estimates for achieving environmental sustainability.

To the knowledge of the task force, no national needs assessment has been conducted for the environment to date; however, such gaps in current analyses do not imply that investment needs are low. On the contrary, results show that investment needs in environmental sustainability are likely high but underscore the difficulty in preparing robust needs assessments (box 6.4).

The task force therefore calls on countries to conduct detailed MDG needs assessments for the environment. International conventions and organizations should offer technical assistance to developing countries that wish to carry out these assessments. In most cases, this will first require the development of needs assessment tools, which should be a top priority for international organizations (see box 6.3).

Box 6.3

Millennium Development Goals needs assessments

Source: UN Millennium Project forthcoming.

Estimating the resources needed to achieve the Millennium Development Goals must be guided by four principles. First, no one size fits all. Country-level assessments are needed, using country-specific coverage data, targets, and unit costs. The UN Millennium Project recommends this approach for all countries preparing MDG-based poverty reduction strategies.

Second, needs assessments must build on a bottom-up estimate of both capital and operating expenditures. The analyses should also quantify human resource and infrastructure requirements for all interventions necessary to meet the Goals.

Third, operationalizing the Monterrey Consensus requires that domestic resource mobilization by governments and households fund as much of the cost of meeting the Goals as possible. Where investment needs exceed domestic resource mobilization, this financing gap must be covered by official development assistance (ODA).

Fourth, increased ODA is necessary to meet the Goals. Donors must make sufficient funds available, with disbursements contingent on the quality of MDG-based poverty reduction strategies and countries' commitment to undertake needed reforms.

Before initiating the UN Millennium Project needs assessment, countries must firmly agree on binding outcome targets to be achieved by 2015, such as lowering nutrient loads in critical freshwater ecosystems or slowing the rate of deforestation to a certain level. Once these outcome targets are set, the needs assessment can begin. The first step is to identify all policies and interventions—defined broadly as provision of services, goods, and infrastructure—needed to meet the specific environment objectives, as described in chapters 4 and 5. The second step is to identify quantitative targets for each intervention to have reached by 2015. The third step is to develop transparent investment models with which to estimate the capital and operating costs of MDG interventions, including human resources and infrastructure. The fourth step is to revise needs estimates iteratively to address the effect of synergies across intervention areas that affect overall investment needs. For example, increased access to modern cooking fuels will lower pressure on forests and other ecosystems, thereby reducing investment needs for reforestation programs. The fifth and final step is to develop a financing strategy by weighing investment needs against substantially increased domestic resource mobilization, to estimate the financing gap.

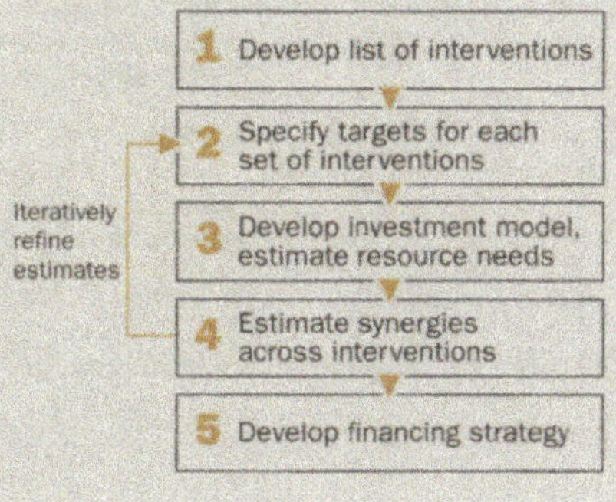

Integrating environmental sustainability into sector strategies

The working group on environmental sustainability must oversee the integration of environmental expertise into the design of sector strategies. Like gender equality, environmental sustainability will remain an elusive goal unless it is integrated into the formulation of sector investment strategies and policies. For

Box 6.4

Costs and benefits of achieving environmental sustainability

Economic analyses of environmental sustainability must quantify both the resources required to meet specific environmental objectives and the damage caused by environmental degradation. While the former is critical for developing sound environmental policies and acquiring the funds to implement them, it is difficult to implement. As a result, few national environment strategies include a rigorous needs assessment.

Several analyses have been carried out to provide indicative cost figures for achieving objectives and targets set in multilateral environmental agreements. The table below compiles selected figures on costing conservation and environmental protection.

Issue	Target or responsible actor	Estimated annual cost ($ billion)	Cost period
Implement Kyoto Protocol	Industrialized countries	49 (under global trading) 108 (under limited trading) 253 (under no trading)	2001–15[a]
Annual losses from climate change if carbon dioxide reaches 550 parts per million	Industrialized countries	304.2 (losses)	
Run a global network of marine protected areas[b]	Covering 20–30% of the seas as recommended	5–19	
Combat desertification	Preventive measures in lands not desertified	2.2–6.7	2001–15
	Preventive and corrective measures in lands moderately desertified	6.0–17.1	
	Prevention, correction, and rehabilitation	15.9–34.6	
Toxic chemicals[c]	Clean up stockpiles of pesticides, prevent new stockpiles	0.25 (estimated)	15-year program
Implement Montreal Protocol	75% incurred by industrialized countries, 25% by developing countries	8.9[d] 17 (benefits)	2001–15
Conserve existing forest areas in developing countries	Hotspots only,[f] 209 million hectares in developing countries	19.6 (for 100%) 7.6 (for 80%) 24.8[e]	2001–15

a. Assumptions are that emissions peak in 2001, reduction in the level of emissions from 2001 to 2015 is linear, and the cost in terms of GDP is proportional to the reduction in emissions.

b. Balmford and others 2004.

c. Buccini and Cortinas 2004.

d. Figure does not include disbursements of the Multilateral Fund, which helps developing countries meet their obligations under the Protocol, estimated at $170 million per year.

e. Zhou 2003, but does not include opportunity costs of logging.

f. Calculations exclude small patches, interior forests (less than 10 kilometers from edge), steep slopes where conversion is unlikely, and currently protected forests; estimate is based on conserving Conservation International's forest hotspots.

Source: World Bank 2002.

this reason, every thematic working group, as well as the planning group, must include technical environment experts, who will identify conflicts between proposed strategies and the environment objectives agreed on by the planning group. As necessary, the experts can draw on the technical expertise of the

Funding for the environment can be made predictable

working group on environmental management to carry out Strategic Environmental Impact Assessments and the recommendations outlined in chapter 5.

Financing environmental sustainability

A viable financing strategy for achieving environmental sustainability is required in all countries, but is currently lacking in most. Such a strategy must satisfy several requirements. First and foremost, it must make more funding available by reallocating resources to the environment, increasing domestic resource mobilization, and raising levels of development assistance where needed. Second, while increased funding will be necessary to implement investments tied directly to slowing or reversing environmental degradation (chapter 4), development assistance directed to sector-specific activities must be conditional on the identification of tradeoffs to the environment and integration of measures to mitigate potentially adverse effects. Third, mechanisms for increasing domestic resource mobilization should support environmental objectives by taxing activities detrimental to the environment and lessening the tax burden on those activities that support the achievement of environmental sustainability, as recommended in chapter 5.

The necessary expansion of the domestic tax base should focus on taxing natural resource consumption, provided that such taxes do not price critical goods and services out of reach of the poor. For example, road and fuel taxes can contribute substantially to national budgets while reducing consumption of fossil fuels. In many cases, a win-win financing strategy might consist of phasing out environmentally damaging subsidies and instituting environmentally friendly ones, creating alternative mechanisms to better address the needs of the poor, and using proceeds to finance investments in environmental sustainability.

By earmarking the proceeds from environmental taxation for environmental expenditures, funding for the environment can be made predictable, thus allowing longer term investments to be planned and implemented. A good example is Brazil's ecological value added tax; a portion of this tax on goods, services, energy, and communications goes to municipalities, based on various indicators of environmental performance (May and others 2002). In Parana and Minas Gerais, this tax has supported an increase in the size and number of protected areas. Similarly, in Colombia, revenues from taxes levied on petroleum and charcoal support activities of the National Environmental System. In Ecuador, a percentage of all electricity revenues are directly allocated to watershed management.

Even with massive increases in domestic resource mobilization, many low-income countries cannot finance all investments in environmental sustainability while simultaneously ensuring the expansion of social services and access to basic infrastructure required to meet the other Goals. These countries therefore require increased external finance in the form of bilateral or multilateral development assistance following the important principles of

Decisionmakers must understand environmental sustainability and monitor progress

shared responsibility and differentiated needs underlying Goal 7, as well as the Monterrey Consensus. In addition, the engagement of the private sector can relieve some of the financial burden from the government (box 6.5).

It is critical that external finance be available for all types of expenditures needed to meet national objectives for environmental sustainability, including operating costs (for example, salaries and consumables). To reduce transaction costs, external finance should be provided in the form of direct budget support or equivalent sectorwide approaches to governments, provided that adequate measures are in place to ensure financial accountability. Civil society organizations in low-income countries will also require substantial financial support to carry out essential activities, as outlined below.

In addition to bilateral development assistance, an array of multilateral channels exist. Chief among them are the Global Environment Facility (GEF), the World Bank's International Development Association (IDA), and the European Development Fund (EDF), as well as regional development banks. The GEF fills an important gap by funding global and regional environmental goods. Available funding is inadequate to implement national MDG-based poverty reduction strategies. In the case of the poorest countries, the IDA, regional development banks, and the EDF must therefore increase financing of environmental sustainability. The bulk of funding to low-income countries must be provided in the form of grants (UN Millennium Project 2005e).

Where possible, countries should strengthen mechanisms to promote private sector activity in support of environmental objectives. One such economic mechanism is the creation of a functioning financial market in which private companies can purchase and sell environmental goods and services. Emissions certificate trading, already implemented in several developed countries, shows encouraging signs of success in mobilizing private resources for environmental sustainability. However, such mechanisms place high demands on public institutions, which must establish and regulate markets effectively. In addition, the success of such mechanisms depends on sophisticated financial markets and major corporate investments in environmental sustainability. These key conditions are not met in most low-income countries, which must therefore depend on less complex mechanisms.

Other promising strategies include funds to support environmentally friendly investments in developing countries. One such example is the Clean Development Mechanism (CDM) established under the Kyoto Protocol (Smith and Scherr 2002). By channeling private capital to technologies geared toward environmental sustainability, the CDM can be a powerful tool for technology transfer.

Monitoring progress

Decisionmakers—and society as a whole—must understand the components of environmental sustainability and monitor progress toward achieving environmental sustainability goals. Indicators are important tools for communicating

Box 6.5

What the private sector can do

As a primary driver of economic activity and the key engine of innovation, the private sector plays a critical role in ensuring environmental sustainability. Many businesses, both individually and in organizations such as the World Business Council for Sustainable Development, emphasize that their long-term interests are tied to environmental sustainability. As key stakeholders in the environment, private sector companies can contribute to sustainable development by developing new technologies and practices for addressing unsustainable patterns of production and consumption; contributing to a strengthened policy regime for environmental sustainability; and actively committing to compliance with environmental and other regulations.

The environmental principles of the UN Global Compact provide an entry point for businesses to address key environmental challenges. These principles are that businesses should support a precautionary approach to environmental challenges, undertake initiatives to promote greater environmental responsibility, and encourage the development and diffusion of environmentally friendly technologies. These three principles direct activity to research, innovation, cooperation, education, and self-regulation, which positively address problems of environmental degradation brought about by human activity.

New tools for environmental sustainability. The private sector's unrivalled capacity to drive technological and organizational innovation must be harnessed more effectively. Thanks to its expertise and dynamism, the private sector will play a critical role in, among other things, developing cheaper technologies that use renewable energy sources and cleaner, more efficient technologies for fossil-fuel energy sources. By designing appropriate incentives and developing markets for such products, government, in turn, can effectively mobilize private enterprises, which want to move into new markets. Companies can work in partnership with the public sector or other agents to deliver needed services and investments for environmental management (UN Millennium Project 2005b).

Moreover, companies can commit to environmentally responsible standards of practice, either individually, as Sony has in its Green Management Strategy,[a] or in consortia, as demonstrated by the commercial banking sector's adoption of the Equator Principles.[b] New models for social entrepreneurship, including fair trade, ecolabeling, and ecotourism, suggest novel strategies for the private sector to support sustainability and underscore its capacity to develop innovative services. Mitigation banks and trading in carbon sequestration are yet other creative tools developed in conjunction with the financial industry.

Policy dialogue and design. The private sector can also advocate for more sustainable policies. Munich Re and Swiss Re, both multinational re-insurance companies, and many other corporations that act on the risks of environmental change, urge governments to implement strong, immediate reductions in greenhouse gas emissions, warning that global climate change will substantially increase the risk of natural disasters and substantially destabilize their profits. These companies are often far ahead of policymakers in recognizing the urgency for action. Governments and civil society should work with them to improve environmental policies. For example, this can include creating new markets for carbon emission permits, which could be operated and traded by private companies who have valuable expertise in the design of new markets.

Similarly, since most environmental regulation is directed at the private sector, private companies can work with governments to lessen the financial burden of environmental regulation without compromising environmental outcomes. Likewise, companies can

(continued on next page)

a. See www.wbcsd.org.

b. See www.equator-principles.com.

Box 6.5
What the private sector can do
(continued)

propose ways in which compliance, as well as enforcement of regulation, can be improved to level the playing field on which all companies depend.

Active compliance. Private sector activities are a major driver of environmental degradation. Small and large companies alike can be part of the problem: extractive industries, heavy manufacturing, small-scale tanneries, or automobile repair shops can each drive the degradation of ecosystems and cause pollution. Much of the responsibility for providing private firms incentives to minimize environmental damage lies with public authorities, who must improve regulatory and enforcement measures, as well as long-term changes in fiscal incentives (chapter 5). Businesses, driven by financial incentives, take a precautionary approach: It is cheaper to invest in prevention than pay for mitigation or compensation. The private sector must commit to strict compliance with government-established rules. Businesses that operate legally, together with consumers, must pressure businesses that are not in compliance. Public accountability and transparency and enlightened self-interest of companies, labor, and civil society are key to success.

scientific and technical information to diverse user groups. The adoption of indicators and their monitoring over time allow countries to establish environmental sustainability goals, measure progress, respond to changes, and make mid-course adjustments to environment and development strategies. Data and indicators can also be used to predict the effects of alternative policy decisions on the environment.

Indicators of progress toward environmental sustainability are needed at global, regional, and national levels to support the monitoring of progress toward meeting environmental objectives and identifying emerging trends and challenges. At the global level, these indicators could be linked directly with the Millennium Development Goals, the World Summit on Sustainable Development 2010 biodiversity targets, and other globally accepted goals. At the national level, they could be linked to the poverty reduction strategies developed for national and local actions where appropriate.

Globally accepted indicators and their systematic application across countries and regions are lacking. Also absent are links between the environment and other sectors. While many countries measure biodiversity and the status of related ecosystems, species threat status, and extent of protected areas, they fail to link these indicators directly to poverty reduction or other Goals. Despite the efforts of researchers, international organizations, governments, and civil society organizations to devise such indicators, few have been identified that are easily monitored and internationally comparable. The indicators suggested for Goal 7 do not capture the complexity of environmental sustainability and are based on available global data rather than key indicators needed to monitor progress toward environmental sustainability.

One problem lies in the extreme diversity and complexity of the environment, which may require hundreds of indicators to map progress across all dimensions and scales. The search for a set of internationally comparable environment indicators of the status of natural capital (ecosystems and their services) using uniform units of measurement should therefore continue under the leadership of international organizations and the secretariats of multilateral environmental agreements. These institutions should support countries that need help in collecting the necessary data and developing reliable indicators. Specific attention should be paid to minimizing data and indicator reporting burdens on countries through harmonization of reporting for multilateral environmental agreements, as well as systematization of reporting on collected data and indicators at the international level.

At the national level, lack of an information culture, coupled with limited national budgets, prevents many countries from devoting sufficient resources to indicator-related work (box 6.6). Even in countries where the importance of having data and indicators is well recognized, it is still a challenge to maintain

Box 6.6
Research institutions and science

As explained in chapter 5, science can greatly enhance the understanding of the environment, while technology can enhance human capacity to manage it. Science and technology therefore provide two fundamental approaches to environmental sustainability. Exactly how these powerful approaches are mobilized, however, remains part of the political process that seeks to reflect broad public interests.

All branches of science—natural, social, political, and physical sciences—can play a productive role in achieving environmental sustainability. Scientists bring unique tools to the table: objective observation, peer review, and modification of approaches based on experience. Subjecting policies to rigorous scientific review often improves their relevance to development needs.

Confronting complex and growing environmental challenges requires research and development that target current problems. Research capacity must be expanded to accelerate progress toward creating better technologies, policies, and practices. Improved capacity is especially needed in developing countries to enable policymakers to assess and chart their own destinies. The task force supports the recommendations prepared by the Task Force on Science, Technology, and Innovation and offers its own here.

First, science and technology institutions must train future engineers, educators, policymakers, and scientists of all kinds. Second, the innovative thinking generated in these institutions must be channeled to provide solutions to development questions. Researchers can develop technological solutions, provide data and information management services, and train local people in solving environmental problems. Given the scale of environmental issues facing humankind today, these contributions have never been more necessary. Third, these institutions can bring together multidisciplinary groups—economists and political, social, and natural scientists—to develop appropriate tradeoff mechanisms and mitigation procedures for situations where a development activity might compromise the ecosystem's ability to provide desirable services. A classic example is water used to produce food in agricultural systems that depletes water for other purposes, such as maintaining healthy aquatic habitats or providing villagers with drinking water year-round. Finally, with the aim of identifying applied research questions, science and technology institutions must be involved in the international coordination of research and assessments.

Improved regional management of the environment is a necessary complement to national strategies

the level of interest and funds required to collect primary data, establish useful indicators, and use them to monitor environmental and sustainable development trends. Countries must generate relevant data and develop indicators that help them frame and implement their policies, as well as monitor policy success. Many countries need both technical and financial assistance to build capacity in primary data collection, data processing and management, and development of integrated databases and data sets and information monitoring systems. While indicators vary by country, they share certain characteristics relating to specificity, measurability, achievability, and temporal scale. Developing a small set of indicators shared by all countries would assist global policymakers and enable cross-country comparisons.

In all countries, environmental data acquisition remains vital to strengthening systems to collect data on the physical environment. Many developing and some developed countries do not systematically monitor key environmental parameters (such as air and water quality, biodiversity, or land degradation), let alone develop indicators from these data for use by decisionmakers and the public. Few developing countries have adequate systems in place to monitor urban air quality or hydrological flows. Many countries therefore must invest in new measurement stations and systems to collect and analyze environmental data. MDG-based poverty reduction strategies must identify these investments and include practical strategies for delivering them.

Equally important is the effort to promote multisectoral and multi-institutional collaboration on data and indicator work—especially among national statistics offices, environment ministries, and other technical institutes—to minimize overlaps and duplication of effort. National statistics offices or environment ministries should coordinate such efforts to ensure that data is freely available and widely shared.

In this context, scientific advisory bodies or offices can play a clear role in providing, analyzing, and translating scientific information for policymakers, students, and the general citizenry. A well informed public can make better decisions at all levels of society and action—from national development planning to individual consumption.

Regional implementation mechanisms

Improved regional management of the environment is a necessary complement to national policies and investments since many environmental challenges extend beyond national borders. Important examples are coastal ecosystem degradation, transboundary watersheds and their accompanying rivers and international lakes, and acid rain.

Regional implementation mechanisms—which often take the form of consultative bodies—that have successfully addressed environmental change include the Mekong River and Nile Basin Initiatives, which have improved joint management of transboundary rivers and watersheds. A promising example

Funding must be earmarked to support long-term regional strategies

is the Amazon Cooperation Treaty Organization, which aims to develop a joint management strategy for the Amazon Basin among all relevant countries. Likewise, the Victoria Basin Commission is a promising start toward coordinating management of East Africa's Lake Victoria.

Several key lessons can be drawn from the limited experience with regional environmental management to date. First, such cooperation requires dedicated regional institutions and concerted investments over long periods of time. Unfortunately, regional institutions tend to lack sufficient operational budgets and the mandate from their member countries that would enable them to pursue their work effectively. Members' contributions to regional organizations are often discretionary and can be withheld at a moment's notice, and regional environmental management remains underfunded. Thus, funding must be earmarked to support the development of long-term regional strategies for environmental management.

Second, effective regional management becomes particularly difficult when the balance between costs and benefits from improved environmental management varies widely across neighboring countries. In situations where upstream countries would incur substantial costs but enjoy few benefits from improved environmental management, substantial resources must be transferred from downstream users. Experience shows that such arrangements are difficult to set up and maintain. In such situations, dedicated financing through such mechanisms as the GEF can be particularly effective in supporting regional environmental management. To this end funding windows for regional investments must be increased by the GEF and other donors.

Third, the diversity of regional challenges and available institutional mechanisms reinforce the lesson that no blueprint exists for strengthening regional cooperation. Indeed, the nature of implementation mechanisms depends critically on the political leadership of individual countries and the extent to which regional organizations can assist. Clearly, regional mechanisms for improved environmental management are neglected and must be strengthened to support country-level, MDG-based poverty reduction strategies.

Global implementation mechanisms

Certain environmental challenges are global and can only be managed through global implementation mechanisms. Examples include climate change, overharvesting of global fisheries, and depletion of the Earth's ozone layer. In addition, certain national-level drivers of environmental degradation, such as international trade in illegally harvested timber or endangered species, require a well coordinated global response to succeed. Thus, issues of global concern require global coordination strongly coupled with regional and national action planning.

Two types of market failures must be addressed to improve global environmental management. First, free access to global resources, such as international fisheries, will lead to overexploitation without coordinated mechanisms

Correcting the market failures requires negotiated international agreements

to control access. Second, in the absence of pricing mechanisms that reflect the true value of environmental goods and services and the true cost of environmental degradation, people's daily decisions will be biased toward higher levels of resource use or pollution and use of inefficient technologies. For example, no effective pricing mechanism exists to incorporate the economic cost of global warming into the prices people pay for emitting greenhouse gases or consuming fossil fuels. As a result of this market failure, individual decisions lead to higher emissions of greenhouse gases than would be the case if the full consequences of their actions were factored into available choices. Hence pricing mechanisms must be introduced that spur research and development of cleaner and more energy efficient technologies.

Since no world government is likely to exist in the foreseeable future, correcting market failures through these and other mechanisms requires negotiated agreements (chapter 5). In the case of global environmental challenges, such agreements must be multilateral since the transaction costs of many bilateral agreements would be excessive.

Like regional policy and institutional strategies, global mechanisms for environmental management must be tailored to each environmental challenge. In addition, they require clear means for reaching decisions and sufficient funding. Unfortunately, the large number of participating countries makes global mechanisms difficult to manage, and they often become unwieldy and ineffective.

To address these challenges, this task force recommends setting up an inter-Convention coordinating mechanism—involving the Convention on Biological Diversity, the Convention to Combat Desertification, the United Nations Framework Convention on Climate Change, and Ramsar—to develop joint programs to find synergies and identify tradeoffs among actions taken under these Conventions. A periodic synthesis (for example, every five years) of the findings of the various international assessments, including the Millennium Ecosystem Assessment, the Intergovernmental Panel on Climate Change (IPCC), and others, would likely reveal those synergies and tradeoffs.

International bodies, including convention secretariats, international organizations, and lending and donor agencies, are already playing an important role in creating change at all levels of society. Agencies are effective in establishing and maintaining partnerships and alliances between and among a variety of sources, including companies and foundations, as well as bilateral and multilateral donors, in advancing the Millennium Development Goals. They are also effective in providing guidance to local and national actors who will implement change on the ground. International convention bodies have defined environmental sustainability principles and standards that many national governments have adopted. Examples include the Guidelines for Sustainable Forest Management of the International Tropical Timber Agreement, the Forest Principles adopted at the United Nations Conference on Environment and Development, and the Cartagena Protocol on Biosafety.

The effectiveness of conventions depends on financial resources and scientific support

The Montreal Protocol has perhaps been singularly effective in reversing degradation of a global environmental resource, the ozone layer. Several unique circumstances contributed to its success, including the rapid availability of economically viable substitutes for ozone-depleting substances and the relatively small number of producing countries. Other conventions that have had considerable success include the Convention on the International Trade in Endangered Species and the Kyoto Protocol of the United Nations Framework Convention on Climate Change (UNFCCC). Despite these examples, overall success in implementing international agreements has been limited.

The effectiveness of conventions depends on sufficient financial resources to support their implementation and access to dedicated financing mechanisms to strengthen operational capabilities. In addition, most conventions must have scientific support to guide national responses. For example, the IPCC of the Climate Change Convention has proven successful in developing sound scientific advice (IPCC 2001), and the reports of the Millennium Ecosystem Assessment close important gaps in the scientific understanding of the natural environment. Moreover, conventions must have a stronger mechanism to monitor and enforce penalties for noncompliance.

All of these international agreements and bodies have excellent work plans and long lists of priority actions, yet few have been implemented. Improving the functioning and effectiveness of existing conventions requires a refocusing on the principle of subsidiarity. This means that decisions should be made at the lowest possible level, to maintain local and national sovereignty while allowing for regional and international decisionmaking. Because of the tremendous coordination challenges involved in reaching global agreements and ensuring implementation, global coordination mechanisms should be used only as a last resort if national or regional implementation is impossible.

Climate change and depletion of the ozone layer are perhaps the two best examples of environmental challenges that require a global coordination mechanism equipped with resources to finance implementation of agreements. As discussed in chapter 4, stabilizing greenhouse gas emissions requires dramatic reductions in projected greenhouse gas emissions that go beyond the commitments made under the Kyoto Protocol of the UNFCCC.

Many other conventions should focus on effectively supporting national interventions to stem environmental degradation. Perhaps the most effective means is to make expertise—scientific knowledge and operational best practices—available to countries. To this end, conventions must strengthen their scientific capacity either through bodies like the IPCC or by acting as a central repository for best practices that countries can access in designing their own strategies. Conventions can also play an important role—as many already do—in advancing the understanding of appropriate indicators that can be used at the country level.

In summary, conventions and international mechanisms supporting their implementation must strengthen their operational expertise and ability to advise interested governments on how to achieve environmental objectives within MDG-based poverty reduction strategies. In many cases, this will require a reorientation away from the normative focus that characterizes many conventions toward enforcement of implementation and a greater ability to design national programs that address environmental challenges at scale.

Notes

Preface

1. "Plan of Implementation of the World Summit on Sustainable Development," the final text of agreements negotiated by governments at the World Summit on Sustainable Development, August 26–September 4, 2002, Johannesburg, South Africa. [www.un.org/esa/sustdev].

Chapter 1

1. Final text of agreements negotiated by governments at the World Summit on Sustainable Development, August 26–September 4, 2002, Johannesburg, South Africa.

2. The Millennium Ecosystem Assessment is a three-year international program designed to meet the needs of decisionmakers and the public for scientific information on the consequences of ecosystem change for human well-being and options for responding to those changes [www.millenniumassessment.org].

Building on the collaboration between the Millennium Ecosystem Assessment process and the UN Millennium Project Task Force on Environmental Sustainability, unpublished material from the Millennium Ecosystem Assessment has been cited in this report with permission, but the material is still in draft form, and some findings and statistics may change prior to publication.

3. Final text of agreements negotiated by governments at the World Summit on Sustainable Development, August 26–September 4, 2002, Johannesburg, South Africa.

Chapter 2

1. The UN considers 50 liters per day per person to be the minimum required for supporting sustainable forms of development and 20 liters per day per person the minimum subsistence requirement.

Chapter 3

1. Adjusted net savings has recently been extended to account for dissavings resulting from soil degradation (Sachs and others 2004).

Chapter 4

1. An *ecosystem* is defined as a dynamic complex of communities of living organisms (including plants, animals, fungi, bacteria, and algae) interacting with the nonliving environment; nonliving elements include air, soil, water, and mineral nutrients.

2. For information on the International Assessment of Agricultural Science and Technology for Development, see www.agassessment.org/index.html.

3. Examples of such international river basins include the Zambezi, Limpopo, Nile, and Niger (Africa); Amazon (South America); and Amur and Mekong (Asia). Examples of national systems that fit the criteria set above are the Krishna, Cauvery, and Yamuna (India); Yangtze (China); and Mississippi and Missouri (United States).

4. A good example is the Danube and Rhine river basins; the Mekong River Commission presents a similar opportunity.

5. More information on the Commission for the Conservation of Antarctic Marine Living Resources management strategy is available at www.ccamlr.org.

6. This target reflects those of the World Summit on Sustainable Development and the Convention of Biological Diversity, which call for a minimum of 70 percent of all exploited fisheries products to be derived from sustainably managed sources by 2010 as a milestone toward the 2015 World Summit on Sustainable Development target.

Chapter 5

1. For further information on the Integrated Water Resources Management Toolbox developed by the Global Water Partnership, see http://gwpforum.netmasters05.netmasters.nl/en/index.html.

2. For further information on the Katoomba Marketplace, see www.katoombagroup.org.

3. See www.intel.com/research/silicon/mooreslaw.htm.

4. See www.ucsusa.org/clean_vehicles/cars_and_suvs/page.cfm?pageID=222.

5. See www.epa.gov/fueleconomy/class-high.htm.

6. See www.iea.org/Textbase/press/pressdetail.asp?PRESS_REL_ID=128.

References

Ali, A. M. S. 2004. "Technological Change in Agriculture and Land Degradation in Bangladesh: A Case Study." *Land Degradation & Development* 15 (3): 283–98.

Bakun, Andrew, and Scarla J. Weeks. 2004. "Greenhouse Gas Buildup, Sardines, Submarine Eruptions and the Possibility of Abrupt Degradation of Intense Marine Upwelling Ecosystems." *Ecology Letters* 7 (11): 1015–23.

Balmford, Andrew, Pippa Gravetock, Neal Hockley, Colin J. McClean, and Callum M. Roberts. 2004. "The Worldwide Costs of Marine Protected Areas." *PNAS (Proceedings of the National Academy of Science of the United States)* 101 (26): 9694–97.

Balmford, Andrew, Aaron Bruner, Philip Cooper, Robert Costanza, Stephen Farber, Rhys E. Green, Martin Jenkins, Paul Jefferiss, Valma Jessamy, Joah Madden, Kat Munro, Norman Myers, Shahid Naeem, Jouni Paavola, Matthew Rayment, Sergio Rosendo, Joan Roughgarden, Kate Trumper, and R. Kerry Turner. 2002. "Economic Reasons for Conserving Wild Nature." *Science* 297 (5583): 950–53.

Balick, Michael J., Elaine Elisabetsky, and Sarah A. Laird, eds. 1996. *Medicinal Resources of the Tropical Forest: Biodiversity and Its Importance to Human Health.* New York: Columbia University Press.

Balvanera, Patricia, and Ravi Prabhu. 2004. "Ecosystem Services: The Basis for Global Survival and Development." Background paper commissioned for the Task Force on Environmental Sustainability. UN Millennium Project, New York.

Barbier, Edward B., and J. C. Burgess. 2001. "The Economics of Tropical Deforestation." *Journal of Economic Surveys* 15 (3): 413–33.

Berkowitz, A. L., C. H. Nilon, and K. S. Hollweg, eds. 2003. *Understanding Urban Ecosystems: A New Frontier for Science and Education.* New York: Springer-Verlag.

Birkhaeuser, Dean, Robert Evenson, and Gershon Feder. 1991. "The Economic Impact of Agricultural Extension: A Review." *Economic Development and Cultural Change* 39 (3): 607–51.

Bojö, Jan, J. Bucknall, K. Hamilton, N. Kishor, C. Kraus, and P. Pillai. 2001. "Environment." In *The Poverty Reduction Strategy Sourcebook.* Volume 1. Washington, D.C.: World Bank.

Bojö, Jan, Kenneth Green, Sunanda Kishore, Sumith Pilapitiya, and Rama Chandra Reddy. 2004. "Environment in Poverty Reduction Strategies and Poverty Reduction Support Credits." Environment Department Report 102. World Bank, Washington, D.C.

Brackett, David, Rodrigo Medellín, Carolina Caceres, and Sue Mainka. 2004. "Biodiversity and Human Livelihoods: The State of the Planet in 2004." Background paper commissioned for the Task Force on Environmental Sustainability. UN Millennium Project, New York.

Bryant, Dirk, Lauretta Burke, John McManus, and Mark Spalding. 1998. *Reefs at Risk: A Map-Based Indicator of Threats to the World's Coral Reefs*. Washington, D.C.: World Resources Institute.

Buccini, John, and Cristina Cortinas. 2004. "Impact of Chemicals Pollution and Use on Health and the Environment." Background paper commissioned for the Task Force on Environmental Sustainability. UN Millennium Project, New York.

Burke, L., Y. Kura, K. Kassem, C. Revenga, M. Spalding, and D. McAllister. 2000. *Pilot Analysis of Global Ecosystems: Coastal Ecosystems*. Technical Report. World Resources Institute, Washington, D.C.

Cincotta, R. P., and R. Engelman. 2000. *Nature's Place: Human Population and the Future of Biological Diversity*. Washington, D.C.: Population Action International. [www.populationaction.org].

Cohen, Mitchell L. 2000. "Changing Patterns of Infectious Disease." *Nature* 406 (6797): 762–67.

Costanza, R., R. D'Arge, R. S. de Groot, S. Farber, M. Grasso, B. Hannon, K. Limburg, S. Naeem, R. V. O'Neill, J. Paruelo, R.G. Raskin, P. Sutton, and M. van den Belt. 1997. "The Value of the World's Ecosystem Services and Natural Capital." *Nature* 387 (6630): 253–60.

Daszak, P., A. A. Cunningham, and A. D. Hyatt. 2000. "Emerging Infectious Diseases of Wildlife: Threats to Biodiversity and Human Health." *Science* 287 (5452): 443–49.

DFID (U.K. Department for International Development), European Commission, UNDP (United Nations Development Programme), and World Bank. 2002. "Linking Poverty Reduction and Environmental Management: Policy Challenges and Opportunities." Working Paper 24824. World Bank, Washington, D.C.

Dilley, Maxx, Robert Chen, Uwe Deichmann, Arthur L. Lerner-Lam, Margaret Arnold, with Jonathan Agwe, Piet Buys, Oddvar Kjekstad, Bradfield Lyon, and Greg Yetman. 2005. *Natural Disaster Hotspots: Global Risk Analysis*. Washington, D.C.: World Bank.

Dulvy, Nicholas K., Yvonne Sadovy, and John D. Reynolds. 2003. "Extinction Vulnerability in Marine Populations." *Fish and Fisheries* 4 (1): 25–64.

Duraiappah, A. 1998. "Poverty and Environmental Degradation: A Review and Analysis of the Nexus." *World Development* 26 (12): 2169–79.

EIA (U.S. Energy Information Administration). 2004a. *International Energy Outlook 2004*. Washington, D.C.: Department of Energy. [www.eia.doe.gov/oiaf/ieo/world.html].

———. 2004b. *World Energy Use and Carbon Dioxide Emissions, 1980–2001*. Washington, D.C. [www.eia.doe.gov/emeu/cabs/carbonemiss/].

Falkenmark, Malin. 2003. *Water Management and Ecosystems: Living with Change*. TEC Report 9. Global Water Partnership, Technical Committee, Stockholm.

FAO (Food and Agriculture Organization of the United Nations). 2001. *Global Forest Resources Assessment 2000*. FAO and National Forestry Action Plans, Rome.

———. 2003. *The State of World Fisheries and Aquaculture 2002*. Rome: FAO. [www.fao.org/documents/show_cdr.asp?url_file=/docrep/005/y7300e/y7300e00.htm].

Ferraro, P. J., and A. Kiss. 2002. "Direct Payments to Conserve Biodiversity." *Science* 298 (5599): 1718–19.

Hamilton, Kirk, and Michael Clemens. 1999. "Genuine Savings Rates in Developing Countries." *World Bank Economic Review* 13 (2): 333–56.

Harvell, C. D., K. Kim, J. M. Burkholder, R. R. Colwell, P. R. Epstein, D. J. Grimes, E. E. Hofmann, E. K. Lipp, A. D. M. E. Osterhaus, R. M. Overstreet, J. W. Porter, G. W. Smith, and G. R. Vasta. 1999. "Emerging Marine Diseases: Climate Links and Anthropogenic Factors." *Science* 285 (5433): 1505–10.

Heal, Geoffrey. 2000. *Nature and the Marketplace.* Washington, D.C.: Island Press.

Hirji, Rafik, and Hans Olav Ibrekk. 2001. "Environmental and Water Resources Management: Input into the World Bank's Environment Strategy and Water Resources Sector Strategy." World Bank, Washington, D.C.

Holdren, J. P., and K. R. Smith. 2000. "Energy, the Environment and Health." In J. P. Holdren and K. R. Smith, eds., *World Energy Assessment: Energy and the Challenge of Sustainability.* New York: United Nations Development Programme.

India, Ministry of Environment and Natural Resources. 2001. *India: National Programme to Combat Desertification.* New Delhi.

IPCC (Intergovernmental Panel on Climate Change). 2001. *Climate Change 2001: Synthesis Report.* Geneva.

Juma, C. 2004. "Technological Innovation and Sustainability: Implementing the Millennium Declaration and the Millennium Development Goals." Paper presented at the High-Level Roundtable on the Implementation of the Millennium Development Goals, November 11, New York.

Kazakhstan government and UN Country Team–Kazakhstan. 2002. *Millennium Development Goals in Kazakhstan.* Almaty.

Kremer, M. 2002. "A Purchase Commitment for Vaccines." In I. Kaul, K. L. Goulven, and M. Schnupf, eds., *Global Public Goods Financing: New Tools for New Challenges, A Policy Dialogue.* New York: United Nations Development Programme.

Lange, Glenn-Marie, Rashid M. Hassan, Kirk Hamilton, and Moortaza Jiwanji. 2003. *Environmental Accounting in Action: Case Studies from Southern Africa.* Cheltenham, U.K.: Edward Elgar Publishing Limited.

Lvovsky, Kseniya. 2001. "Health and Environment." Environment Strategy Background Paper 1. World Bank, Washington, D.C.

Masters, William. 2002. "Research Prizes: A Mechanism to Reward Agricultural Innovation in Low-Income Regions." *AgBioForum* 5 (4): 1–5.

Matthews, Emily, Richard Payne, Mark Rohweder, and Siobhan Murray. 2000. *Pilot Analysis of Global Ecosystems: Forest Ecosystems.* Washington, D.C.: World Resources Institute.

May, Peter, Fernando Veiga Nieto, Valdir Denardin, and Wilson Loureiro. 2002. "Using Fiscal Instruments to Encourage Conservation: Municipal Responses to the Ecological Value-Added Tax in Parana and Minas Gerais, Brazil." In J. Bishop, S. Pagiola and N. Landele-Miles, eds., *Selling Forest Environment Services.* London: Earthscan.

McNeely, Jeffrey, and Patrick Milimo. 2004. "Policy Interventions for Environmental Sustainability." Background paper commissioned for the Task Force on Environmental Sustainability. UN Millennium Project, New York.

Millennium Ecosystem Assessment. 2003. *Ecosystems and Human Well-being: A Framework for Assessment.* Washington, D.C.: Island Press.

Millennium Ecosystem Assessment. 2004a. *Condition and Trends Assessment Report.* 2nd review draft. To be published in 2005 by Island Press, Washington, D.C. [Building on the collaboration between the Millennium Ecosystem Assessment process and the UN Millennium Project Task Force on Environmental Sustainability, unpublished material from the Millennium Ecosystem Assessment has been cited in this report with permission, but the material is still in draft form, and some findings and statistics may change prior to publication.]

———. 2004b. *Scenarios Report.* 2nd review draft. To be published in 2005 by Island Press, Washington, D.C. [See note for 2004a.]

———. 2004c. *Responses Assessment Report.* 2nd review draft. To be published in 2005 by Island Press, Washington, D.C. [See note for 2004a.]

Myers, R. A., and B. Worm. 2003. "Rapid Worldwide Depletion of Predatory Fish Communities." *Nature* 423 (6937): 280–83.

OECD (Organisation for Economic Co-operation and Development). 2004. *Implementing Sustainable Development: Key Results 2001–2004.* Paris.

Oldeman, L. R., R. T. A. Hakkeling, and W. G. Sombroek. 1991. *World Map of the Status of Human-Induced Soil Degradation: An Explanatory Note.* Second Revised Edition. Wageningen, Netherlands, and Nairobi: International Soil Reference and Information Centre and United Nations Environmental Program.

Oldeman, L. R., V. W. P. van Engelen, and J. H. M. Pulles. 1991. "The Extent of Human-Induced Soil Degradation: An Appendix." In L. R. Oldeman, R. T. A. Hakkeling, and W. G. Sombroek, eds., *World Map of the Status of Human-Induced Soil Degradation: An Explanatory Note.* Wageningen, Netherlands, and Nairobi: International Soil Reference and Information Centre (ISRIC) and United Nations Environment Program (UNEP).

Oxfam. 2004. *From 'Donorship' to 'Ownership'? Moving Towards PRSP Round Two.* Briefing Paper 51. Oxford, U.K. [www.oxfam.org/eng/pdfs/pp040119_prsp_.pdf].

Pandita, Sanjiv. 2004. "Civil Society and Governance: Case Study on Right to Clean Air Campaign." Society for Participatory Research in Asia. New Delhi. [www.ids.ac.uk/ids/civsoc/final/india/ind6.doc].

Partnership for Principle 10. 2004. "Report: Second Annual Meeting of the Partnership for Principle 10 Committee of the Whole, June 23–24, 2004." Washington, D.C. [www.pp10.org/PP10_COW_report_2004.pdf].

Patz, J. A., and N. D. Wolfe. 2002. "Global Ecological Change and Human Health." In A. Aguirre, R. Ostfeld, G. Tabor, D. House, and M. Pearl, eds., *Conservation Medicine: Ecological Health in Practice.* Oxford, U.K.: Oxford University Press.

Patz, J. A., P. Daszak, G. M. Tabor, A. A. Aguirre, M. Pearl, J. Epstein, N. D. Wolfe, A. M. Kilpatrick, J. Foufopoulos, D. Molyneux, D. J. Bradley, and Members of the Working Group on Land Use Change and Disease Emergence. 2004. "Unhealthy Landscapes: Policy Recommendations on Land Use Change and Disease Emergence." *Environmental Health Perspective* 112: 1092–98.

Pauly, D., V. Christensen, J. Dalsgaard, R. Froese, and F. Torres Jr. 1998. "Fishing Down Marine Food Webs." *Science* 279 (5352): 860–63.

Pauly, Daniel, Villy Christensen, Sylvie Guenette, Tony J. Pitcher, U. Rashid Sumaila, Carl J. Walters, R. Watson, and Dirk Zeller. 2002. "Towards Sustainability in World Fisheries." *Nature* 418 (6898): 689–95.

Pearce, David, and Edward B. Barbier. 2000. *Blueprint for a Sustainable Economy.* London: Earthscan.

Pearl, Mary, Damayanti Buchori, and Suzana Padua. 2004. "Human and Institutional Capacity Building through Education and Training." Background paper commissioned for the Task Force on Environmental Sustainability. UN Millennium Project, New York.

Pikitch, E. K., C. Santora, E. A. Babcock, A. Bakun, R. Bonfil, D. O. Conover, P. Dayton, P. Doukakis, D. Fluharty, B. Heneman, E. D. Houde, J. Link, P. A. Livingston, M. Mangel, M. K. McAllister, J. Pope, and K. J. Sainsbury. 2004. "Ecosystem-based Fishery Management." *Science* 305: 346–47.

Repetto, Robert. 2004. "Economic Policy Interventions for Sustainable Development and Nature Protection." Background paper commissioned for the Task Force on Environmental Sustainability. UN Millennium Project, New York.

Roberts, Callum M., Colin J. McClean, John E. N. Veron, Julie P. Hawkins, Gerald R. Allen, Don E. McAllister, Cristina G. Mittermeier, Frederick W. Schueler, Mark Spalding, Fred Wells, Carly Vynne, and Timothy B. Werner. 2002. "Marine Biodiversity Hotspots and Conservation Priorities for Tropical Reefs." *Science* 295 (5558):1280–84.

Sachs, J. D., J. McArthur, G. Schmidt-Traub, M. Kruk, C. Bahadur, M. Faye, and G. McCord. 2004. "Ending Africa's Poverty Trap." *Brookings Papers on Economic Activity* 2: 117–216.

Sanchez, P. 2002. "Soil Fertility and Hunger in Africa." *Science* 295 (5562): 2019–20.

Scherr, Sara J., Andy White, and David Kaimowitz. 2003. *A New Agenda for Forest Conservation and Poverty Alleviation: Making Markets Work for Low-Income Producers.* Forest Trends and the Center for International Forestry Research, Washington, D.C.

Secretariat of the Convention on Biological Diversity. 2001. *Global Biodiversity Outlook 2001.* Montreal.

Shepherd, Gill. 2004. *The Ecosystem Approach: Five Steps to Implementation.* Gland, Switzerland, and Cambridge, U.K.: World Conservation Union.

Smith, Joyotee, and Sara J. Scherr. 2002. *Forest Carbon and Local Livelihoods: Assessment of Opportunities and Policy Recommendations. CIFOR Occassional Paper 37.* Center for International Forestry Research, Bogor Barat, Indonesia.

Thomas, C. D., A. Cameron, R. E. Green, M. Bakkenes, L. J. Beaumont, Y. C. Collingham, B. F. N. Erasmus, M. Ferreira de Siqueira, A. Grainger, L. Hannah, L. Hughes, B. Huntley, A. S. van Jaarsveld, G. F. Midgley, L. Miles, M. A. Ortega-Huerta, A. Townsend Peterson, O. L. Phillips, and S. Williams. 2004. "Extinction Risk from Climate Change." *Nature* 427 (6970): 145–48.

UNDESA (United Nations Department of Economic and Social Affairs) Population Division. 2004. *Urban and Rural Areas 2003.* New York.

UNDP (United Nations Development Programme). 2002. *Poverty and Environment Initiative.* New York.

———. 2003a. *Human Development Report 2003.* New York: Oxford University Press.

———. 2003b. *The Kyrgyz Republic: Millennium Development Goals Progress Report.* UNDP Kyrgyzstan, Bishkek.

———. 2003c. *Tajikistan Millennium Development Goals Progress Report.* UNDP Tajikistan, Dashanub.

UNEP (United Nations Environment Programme). 2001. *Asia-Pacific Environment Outlook 2.* Pathumthani, Thailand: UNEP Regional Resource Center for Asia and the Pacific.

———. 2003. *GEO Global Environmental Outlook 3.* New York. [www.unep.org/geo/geo3/].

———. 2004. *Global Environmental Outlook 2003 Year Book.* New York. [www.unep.org/GEO/yearbook/].

UNESCAP (United Nations Economic and Social Commission for Asia and the Pacific). 2003. *Promoting the Millennium Development Goals in Asia and the Pacific: Meeting the Challenges of Poverty Reduction.* Bangkok.

UNESCO–WWAP (United Nations Scientific, Educational and Cultural Organization–World Water Assessment Program). 2003. *World Water Development Report 2003: Water for People—Water for Life.* Paris.

UNFPA (United Nations Population Fund). 2001. *The State of World Population 2001, Footnotes and Milestones: Population and Environmental Change.* New York. [www.unfpa.org/swp/2001/english].

United Nations, European Commission, IMF (International Monetary Fund), OECD (Organisation for Economic Co-operation and Development), and World Bank. 2003. *Handbook of National Accounting: Integrated Environmental and Economic Accounting.* Washington, D.C. [http://unstats.un.org/unsd/envAccounting/seea2003.pdf].

UN Millennium Project. 2005a. *Health, Dignity, and Development: What Will It Take?* Report of the Task Force on Water and Sanitation. London: Earthscan.

———. 2005b. *Innovation: Applying Knowledge in Development.* Report of the Task Force on Science, Technology, and Innovation. London: Earthscan.

———. 2005c. *Taking Action: Achieving Gender Equality and Empowering Women.* Report of the Task Force on Education and Gender Equality. London: Earthscan.

———. 2005d. *Who's Got the Power? Transforming Health Systems for Women and Children.* Report of the Task Force on Child Health and Maternal Health. London: Earthscan.

———. 2005e. *Investing in Development: A Practical Plan to Achieve the Millennium Development Goals.* London: Earthscan.

———. 2005f. *A Home in the City.* Report of the Task Force on Improving the Lives of Slum Dwellers. London: Earthscan.

———. 2005g. *Halving Hunger: It Can Be Done.* Report of the Task Force on Hunger. London: Earthscan.

———. Forthcoming. *Handbook for MDG Best Practice.* New York.

UNU (UN University)/PLEC (People, Land Management, and Ecosystem Conservation). Website. [www.unu.edu/env/plec].

Warwick, Hugh, and Alison Doig. 2004. *Smoke: The Killer in the Kitchen: Indoor Air Pollution in Developing Countries.* London: ITDG Publishing.

Watson, Robert T. 2004. "Energy and Environment." Background paper commissioned by the Task Force on Environmental Sustainability. UN Millennium Project, New York.

WCED (World Commission on Environment and Development). 1987. *Our Common Future.* [The Brundtland Report]. New York: Oxford University Press.

WHO (World Health Organization). 1997. "Health and Environment in Sustainable Development: Five Years after the Earth Summit." WHO/EHG/97.8. Geneva.

———. 2000. *Air Quality Guidelines for Europe.* Second Edition. WHO Regional Publications: European Series 91. Copenhagen: WHO Regional Office for Europe.

WHO (World Health Organization) and UNICEF (United Nations Children's Fund). 2000. *Global Water Supply and Sanitation Assessment 2000 Report.* WHO. Geneva. [www.who.int/docstore/water_sanitation_health/Globassessment/GlobalTOC.htm].

Wood, S., K. Sebastian, and S. Scherr. 2000. *Pilot Analysis of Global Ecosystems: Agroecosystems.* Washington, D.C.: World Resources Institute and International Food Policy Research Institute.

World Bank. 1997. *Five Years after Rio: Innovations in Environmental Policy.* Environment Sector Report 16703. Washington, D.C.

———. 2002. "Costing the 7th Millennium Development Goal: Ensure Environmental Sustainability." Washington, D.C.

———. 2004a. *World Development Indicators 2004.* Washington, D.C.

———. 2004b. *Sector Brief: Environment Sector in Middle East and Northern Africa.* Middle East and North Africa Region, Washington, D.C. [www.worldbank.org].

World Economic Forum, Yale Center for Environmental Law and Policy, and CIESIN (Center for International Earth Science Information Network). 2002. *Environmental Sustainability Index.* [www.ciesin.columbia.edu/indicators/ESI].

World Water Commission. 2000. *Commission Report: A Water Secure World.* Cairo: World Water Council.

WRI (World Resources Institute). 1998. *World Resources: 1998–99.* New York: World Resources Institute, United Nations Environment Programme, United Nations Development Programme, World Bank, Oxford University Press.

Zhou, Liying. 2003. "An Estimation of Global Biodiversity Conservation Costs." World Bank, Washington, D.C.

The UN Millennium Project is an independent advisory body commissioned by the UN Secretary-General to propose the best strategies for meeting the Millennium Development Goals (MDGs). The MDGs are the world's targets for dramatically reducing extreme poverty in its many dimensions by 2015—income poverty, hunger, disease, exclusion, lack of infrastructure and shelter—while promoting gender equality, education, health, and environmental sustainability.

The UN Millennium Project is directed by Professor Jeffrey D. Sachs, Special Advisor to the Secretary-General on the Millennium Development Goals. The bulk of its analytical work has been carried out by 10 thematic task forces comprising more than 250 experts from around the world, including scientists, development practitioners, parliamentarians, policymakers, and representatives from civil society, UN agencies, the World Bank, the International Monetary Fund, and the private sector. The UN Millennium Project reports directly to UN Secretary-General Kofi Annan and United Nations Development Programme Administrator Mark Malloch Brown, in his capacity as Chair of the UN Development Group.

Task Force on Hunger
Halving hunger: it can be done

Task Force on Education and Gender Equality
Toward universal primary education: investments, incentives, and institutions

Task Force on Education and Gender Equality
Taking action: achieving gender equality and empowering women

Task Force on Child Health and Maternal Health
Who's got the power? Transforming health systems for women and children

Task Force on HIV/AIDS, Malaria, TB, and Access to Essential Medicines
Working Group on HIV/AIDS
Combating AIDS in the developing world

Task Force on HIV/AIDS, Malaria, TB, and Access to Essential Medicines
Working Group on Malaria
Coming to grips with malaria in the new millennium

Task Force on HIV/AIDS, Malaria, TB, and Access to Essential Medicines
Working Group on TB
Investing in strategies to reverse the global incidence of TB

Task Force on HIV/AIDS, Malaria, TB, and Access to Essential Medicines
Working Group on Access to Essential Medicines
Prescription for healthy development: increasing access to medicines

Task Force on Environmental Sustainability
Environment and human well-being: a practical strategy

Task Force on Water and Sanitation
Health, dignity, and development: what will it take?

Task Force on Improving the Lives of Slum Dwellers
A home in the city

Task Force on Trade
Trade for development

Task Force on Science, Technology, and Innovation
Innovation: applying knowledge in development

For Product Safety Concerns and Information please contact our EU
representative GPSR@taylorandfrancis.com
Taylor & Francis Verlag GmbH, Kaufingerstraße 24, 80331 München, Germany

www.ingramcontent.com/pod-product-compliance
Lightning Source LLC
LaVergne TN
LVHW080853300926
849162LV00003B/413
* 9 7 8 1 1 3 8 4 7 1 8 2 5 *